# 群菁汇
高端售楼会所大赏
Essence Extracts
Top Sales Clubs

深圳市创扬文化传播有限公司 策划　徐宾宾 编

凤凰出版传媒集团 ｜ 凤凰空间
江苏人民出版社 ｜ IFENGSPACE

PREFACE 序言

作为一名设计师,最大的成就感莫过于看到客户受益于自己的设计成果,也就是说自己对设计做出的贡献,帮客户解决了问题,成就感也油然而生。

在从事多年设计的过程中,售楼部这些设计项目渐渐成为了设计师的工作重点。其实售楼部与我们日常生活中的关系并不紧密,说白了就是楼盘的专卖店,但里面却综合了各种各样的空间,并且楼盘的气质要与售楼部气质一致。因此,一个好的售楼部的标准,不仅仅要好看,有创意,而且还要看其是否真正地符合营销要求,从而推动销售。

售楼部的展示性与营销息息相关,除了传达产品信息等各方面基本功能外,设计师通过对空间环境和气氛的塑造手法,把消费者的视觉、听觉,甚至味觉调动起来,提高他们了解商品的兴致,尽可能达到提高销售的目的。要多了解楼盘目标客户群的年龄层次、爱好,以及消费倾向等方面,这样对项目的理解认知和把握程度会更深,以减少不协调的现象。

在定位清晰后,设计师会从多方面考虑楼盘的品牌信息,如结合当地气候,通过色调来进行搭配,在寒冬里传达暖意;在豪宅社区的售楼部尽量把空间做大,挑高要高,把豪宅的气派与格调带出来;针对消费者比较年轻的,设计可以表现得酷一点,尽量用多一些新型材料与设计手法来表达。

有些售楼部空间规划如出一辙,接待台直进入口处,背景墙上大大的LOGO,再进入大厅、模型区。这样的设计很常规,不会出错,但也不会出彩,适合一般住宅产品。如挑选别墅、豪宅等产品的客户,则会考虑此产品定位是否符合自己的身份,在这时候,售楼部的设计往往就要突破常规设计,从多角度去探讨这类设计,如做成类似艺术馆形式的,很新颖科幻的,很富有东方情调的,很度假休闲的……当售楼部的内涵越丰富,它对客户的吸引力就越高。

这样结合项目的地理、人文、审美习惯、项目定位等多个要素来考虑,才可能做出好设计。

在未来的售楼部设计中,作为设计师更希望能多注重"人"的因素,能更亲和、更实用,而不再盲目追逐哗众取宠。把握好设计的尺度,在完成作品的同时满足客户种种美与功能上的需求。

王小锋
广州尚逸装饰设计有限公司

As a designer, the biggest achievement is seeing that the customers can benefit from my design. That is, if we make contribution to the design career and help customers to solve the problems, sense of achievement will arise spontaneously.

During the process of many years' design, the design projects of houses' sales center are becoming our key emphasis in work gradually. In fact, houses' sales centers don't connected closely to our daily life. Frankly speaking, they're just specialty stores for selling houses. But it combines all kinds of spaces, and its character should be in the same line with that of the buildings on sale. Therefore, the standard of a good houses' sales center should be not only with good-looking and innovation, but also be in accordance with the sales requirement and be the impetus of sales.

The display is closely tied up with sales. Besides conveying the basic functions of products information, we make good use of consumers' vision, audio, and even taste sense, which can increase their interest on knowing the goods and increase the sales accordingly. Knowing the target customers' age, hobby, and consumption tendency etc., can help to understand further about the projects and avoid being inharmonious.

After positioning clearly, we will consider the brand information of buildings. For example, considering local climate, we use colors to match, and convey the warmth in winter; Enlarge the inner space of the sales center in luxury community and make sure the floor height be high, which can show perfectly the style and pattern of luxury houses; Or design in a cool way for younger consumers, and use more new types of material and designing ways.

The layouts of the space in some sales center follow a similar way. The reception desk is at the entrance, big logo on the background wall, then hall and mould area. This design is very regular, and it won't make mistake but won't stand out, which is only suitable for common buildings. The customers who are choosing villas and luxury houses will consider if the positioning of this product fit for their position. In this situation, the design of the sales center should break through the common rules and probe from different aspects, e.g. design the form of art museum, novelty and fiction, eastern romance, leisure and recreation...The more contents the houses' sales center has, the more attractive it will be for the customers.

Designs can only be nice when combining the factors such as geography, culture, aesthetic appreciation, positioning of the projects.

In the design of future houses' sales center, as a designer, I do hope that the attention can be more focused on the element of "human", which can be more compatible, more applied, but not play to the gallery. When working on the design, we should control its scale and meet the customers' various needs about beauty and function.

Xiaofeng Wang
Guangzhou Shangyi Decoration&Design Co.,Ltd.

CONTENTS 目录

沈阳保利康桥会所 Shenyang Baoli Kangqiao Club	006	
保利西海岸英式会所 Baoli West Coast English-style Club	028	
贡山一号会所 Gong Mountain No.1 Club	046	
乌兹别克斯坦国际论坛宫殿 Palace of the International Forums Uzbekistan	056	
沈阳保利十二橡树庄园别墅会所 Shenyang Baoli Twelve Oak Manor Club	088	
远雄新都 Farglory New City	104	
早安清境接待会馆 Qingjing Reception Hall	116	
如意会所 The Ideal Club	132	
上海华屋馆 Shanghai Huawu Club	140	
女王的魔方 Queen's Cube	148	
济南他山会所 Jinan Other Mountains Club	154	
160	巴德拉格斯酒店会所 Grand Resort Club Bad Ragaz	
174	济南未来城会所 Jinan Future City Club	
180	宾格台球 Binge Billiards	
190	海华国际之星会所 Haihua International Star Club	
200	南宁荣和山水绿城会所 Nanning Ronghe Happy House	
208	美的容桂御海东郡会所 Ronggui Royal Sea Club	
216	深圳万科棠樾会所 Shenzhen Vanke Tangyue Club	
224	富临名家昆仑商务会所 Felicity Famous Kunlun Business Club	
234	仁发建设三峰接待中心 Sam-fong Renfa Constructions Reception Center	
242	顺德海岸星座会所 Shunde Coast Constellation Club	
246	北京像素售楼处 Beijing Xiangsu Property Broker	

成都蓝光和骏47亩售楼中心 Chengdu Languang Hejun 47 Mu Sales Center	254	324	杭州金都高尔夫销售处 Hangzhou Jindu Golf Sales Center
东方金石接待中心 Oriental Jinshi Reception Center	258	328	中海锦城会所 Zhonghai Jincheng Club
陆江当代 Green Generation	266	332	汇通国际公寓售楼部 Huitong International Apartment Sales Department
HYL Gallery自主建筑创志馆 HYL Gallary Independent Architecture	272	336	珠光新城销售中心 Pearl Park Sales Center
森林苑 Forest Court	280	342	富通地产天邑湾售楼处 Tianyi Bay Sales Office of Futong Real Estate
绿城房产诸暨售楼部 Greentown Real Estate Zhuji Sales Department	288	348	成都中信未来城售楼中心 Zhongxin Future City Sales Center
广州创逸雅苑售楼中心 Guangzhou Chuangyi Yayuan Sales Center	298	356	中茵上城国际售楼部 Zhongyin Uptown International Sales Center
杭州大家武林府售楼中心 Wulinfu Sales Center	304	364	沈阳亚美利加售楼中心 Shenyang Yameilijia Sales Center
沈阳保利心语花园售楼处 Shenyang Baoli Xinyu Garden Sales Office	308		
武汉福星国际城售楼处 Fuxing International Sales Offices in Wuhan City	316		
中海金沙湾东区售楼中心 Zhonghai Jinsha Bay Sales Center	320		

Shenyang Baoli Kangqiao Club

沈阳保利康桥会所

Baoli kangqiao club is located on the first water bank of Huihe north coast and Wulihe Park. The total coverage is 109 thousand square meters. The building adopts classic architectural style of Art Deco, which has a distinctive conception that reaches top standard in Shengyang. Besides, the garden as well utilizes Art Deco style to arrange theme background, presenting a hundred-year classic elegant living atmosphere in full aspects.

设 计 师：王赟、王小峰
设计公司：广州尚逸装饰设计有限公司
建筑面积：1200平方米

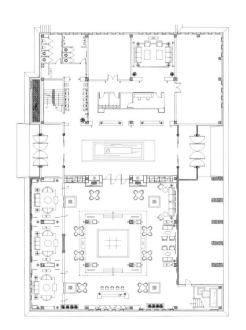

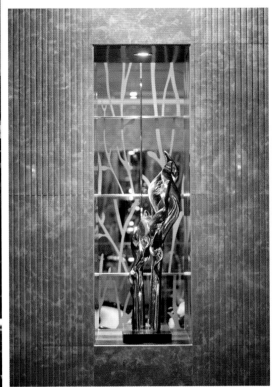

保利康桥位于浑河北岸、五里河公园一线水岸处。会所总建筑面积1200平方米，楼体采用Art Deco经典建筑风格，理念鲜明，达到沈阳顶级标准。园区内Art Deco风格规划主题景，全景式展示百年经典的优雅人居风情。

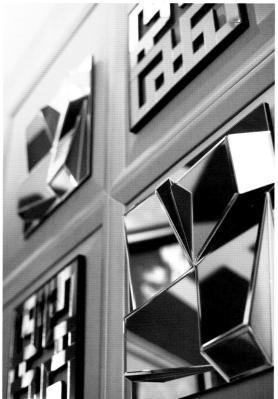

Baoli West Coast English-style Club
保利西海岸英式会所

The design case is located on north Cai Bing road in the northern of Golden Sand land. It is east to Bing Jiang Park which is 100 meters wide. Overlooking from the house, the owner can enjoy the upstream scenery of Pearl River. It is also bordering west to Xun mountain range with a length of 2.5 kilometers and an area of more than 7 thousand acre, which is born as a "natural oxygen bar" and also is full of nature resource. The Baoli Club adopts a rare England tudor architecture style, whose environment is elegant, tranquil. We make most of the physical feature of that place to establish a dimensional nature garden with England style. With a unique vision of international villa, we present the remarkable mansion in Guangzhou , which also is an masterpiece of Baoli estate for the last eighteen years.

设计师：王赟、王小峰
设计公司：广州尚逸装饰设计有限公司
建筑面积：1100平方米

The mansion is the key part of the estate. Its existence not only is for commercial purpose but also an expression of estate businessmen. Therefore, we define with two roles as sales department and private garden in England style that is showing the lifestyle of England royal family.

The case is focusing on a single building with three storeys. The underground floor is created as lobby, reception, wine cellar, reading center and projection room; the first floor has bar, VIP conference room, and cigar room; the second floor is formed by dining hall and work offices, etc. We introduce a sort of England classic rejuvenation into interior design, then use a modern approach to make a design model that is practical but without losing fashion elegance.

We prefer a rational and intense layout for the space and we spare no effort to create the uniqueness of every single space room, which presents a united style of the whole design, also every unique idea tip can be seen in each region.

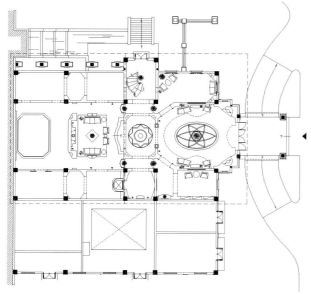

Walking into the lobby, people can see the lustrous ceiling light which reflects the whole space spectacular and elegant. Besides, we bring European style furniture and sculpture into interior design. The employment of lighting ornaments and marble material enable the lobby to pay more attention to the variety of material and space integration. The design conception of this estate is clearly expressed by the highlighting of level and depth.

We put a cozy couch in the reception room. In addition, the smooth light beam, red color for the wall and the well decorated fireplace, all of them send an elegant and warm message to every potential buyer.

The club also installs with solemn reading room and appealing, individual wine cellar; the bar is built to hold some activities or parties, which is classic but not lacking modern sense; the cigar room is fit for businessmen who have a real taste of quality life; the dining hall can accommodate lots of people for conference, generous and stable. So the club embraces plural practical region functions that can satisfy different buyers' life needs.

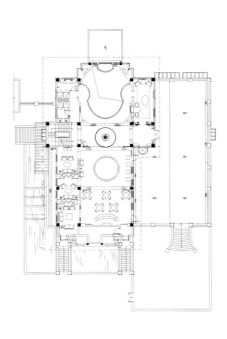

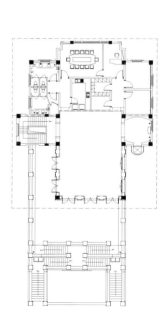

项目位于金沙洲北部彩滨北路上,东临100米宽的滨江公园,相隔眺望珠江上游江岸的景色,西邻467公顷延绵2.5千米山脉的浔峰山天然"大氧吧",有着非常丰富的自然资源。保利西海岸采用市场较为少见的英伦都铎建筑风格。地域环境优美、静谧,充分利用地势打造一个英式自然立体园林。以其独特的国际豪宅视野极力打造广州标致性豪宅,是保利地产18年来又一巅峰巨献。

会所作为整个楼盘的重要组成部分,不仅仅是为其商业目的而存在,更要体现房地产商的品位。因此,我们把它定位于一个售楼部和英伦私家庄园两用来考虑,体现了英式贵族的生活方式。

本项目为三层单体建筑,负一层为大堂、接待处、酒窖、书吧、投影室;一层为酒吧、VIP洽谈区、雪茄房;二层为宴会厅、办公区等。把英伦这种古典复兴的特点引入室内设计中,用现代的处理手法打造成既重实用又不失时尚典雅的设计典范。

在空间布局上,合理紧凑,致力打造每个独立空间的塑造和英式精致时尚的风格,风格统一而又各具特点的设计心思体现在每个区域。

步入大堂,璀璨的水晶吊灯将整个大堂映射得典雅华丽,欧式家具、雕塑等将人们带进了室内陈设和装饰之中,灯饰、石材的运用,使大堂装饰更讲究材质的变化和空间的整体性。强调层次和深度,明确地表现出楼盘的设计概念和风格。

接待处优雅舒适的沙发,柔和的灯光,配合红色的墙身,富有装饰性的壁炉等,给每位到来的顾客一种典雅、细腻的感觉。

会所还设有庄重肃穆的书吧;具有别致情调的个性小酒窖;适合举办活动、派对,风格经典而又不失时尚感的酒吧;适合懂得品味高尚生活的商务人士的雪茄房;可举办容纳多人的商务洽谈,风格大气、稳重的宴会厅等。多元化的区域功能,迎合每一位买家不同的生活需求。

Gong Mountain No.1 Club

贡山一号会所

The project, located in Qinggang forest area in Zigong City, is one of the most important housing estates in Gongjing District. Enjoying an ideal geographic location and surrounded by top-grade residential area, villas and a number of ancillary commercial facilities, this project aims to create a stylish, high-quality, practical and comfortable space, where people can have a new experience and awareness, leading people to the pursuit of luxury homes.

The designer fixes the style as "New Deco style" in an attempt to interpret European traditional elements in modern technique of expression. A magnificent but affinity space is presented through various fashionable totems, updated traditional vocabularies, and personalized contemporary art.

Through the hall, you come into the space. The first thing coming into your eyes is a large blue-violet carpet with many symbolic language. Two plum-shaped pillars carved from black and white stones greet you in each

设 计 师：刘卫军
设计公司：PINKI（品伊）创意集团&美国
　　　　　IARI刘卫军设计师事务所
项目地点：中国四川
建筑面积：2500平方米
主要建材：米黄大理石、皮革、木饰面、壁
　　　　　纸、铁艺

side, and artfully guide your attention to the fifteen-meter-high dome of the lobby. The whole looks magnificent. Luxuriant decoration, organic sculptures, distorting mirror full of a sense of sequence, a strong color contrast, and geometric shapes with dissociation and conversion, all add an extra touch of fashion and elegance to the space.

The light belts of the ceiling well connect the lobby with the negotiating area, which strengthen the feeling of depth of the space and make space more interesting. The display area for models and the wine bar mainly take the design of simplicity and pureness. The tall door guard lines, vertical wood columns, and huge rectangular chandeliers make the space appear to be wide comfortable. The great skylight modifies the feeling of depth of the space.

The planning of the overall layout takes the principles of openness and spaciousness. The space looks both enclosed and transparent, and is connected together as a whole through the details. Delicate and beautiful lamps, metal arts, and elegant furniture make the air of classic and modern to the extreme, and also give the temperature of the times and cultural conservation to the space.

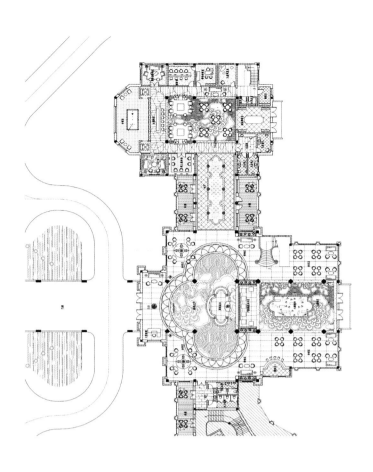

该项目位于自贡市青杠林片区，是贡井区重点打造的居住片区之一，地理位置优越，周边是高档小区、别墅以及一些附属商业设施。该项目本着营造时尚、高品位、实用的舒适空间，使人们在这里能有全新的感受和认知，引领人们对奢华居所的追求。

设计师把风格定位为"新装饰主义风格"，试图以现代的表现手法将欧洲传统元素进行演绎，通过把各种时尚"图腾"元素、更新的传统语汇、个性化的当代艺术有机结合，展现出宏伟而又亲和的空间。

进入空间，经过门厅的缓冲，首先步入眼帘的是大块面并且富有符号语言的蓝紫色地毯，两列黑白根石材雕琢而成的梅花形柱子列阵相迎，巧妙地把人们的视线指引向十五米高的大堂穹顶，显得宏伟壮丽。富丽的装饰，有机体的雕塑，序列感的哈哈镜、强烈的色彩对比、几何性造型经过转换和抽离，更增添几分时尚与优雅。

天花灯带把大堂与洽谈区有机地联合起来，加强了空间的纵深感，使空间更具有趣味性。模型展示区、红酒吧区的表现以简洁、纯粹为主，高大的门套线、纵向的木珊、大体量的矩形吊灯，使空间显得开阔舒适，恢弘的天窗更是修饰了空间的纵深感……

在整体布局规划上，采用开放宽敞原则，空间区域之间既围合又显通透，并以设计的细节将其串连为一体。精致、唯美的灯具，金属艺术品，典雅的家具……把经典与现代气息推向了极致，并赋予空间时代的温度与文化的涵养。

Palace of International Forums
Uzbekistan
乌兹别克斯坦国际论坛宫殿

The Palace of International Forums (Uzbekistan) stands on Amir Timur Square in the very centre of Tashkent. The country's most important representative building is designed as a platform for hosting acts of state, congresses, conferences and other cultural highlights. Our task was to give the interior a contemporary form, while incorporating elements from traditional Uzbek architecture. The result is a cosmopolitan, communicative interior clothed in exclusive materials. Planar ornamentation, organic movement, crystals, precious metals and a fascinating interplay of artificial and natural light all become a source of inspiration.

FOYER

Classical in its external appearance, the building prepares for the modernity within through the extensively

设计师：Peter Ippolito、Gunter Fleitz、Steffen Ringler、Tilla Gold berg、Silke Schreier、Swetla na Wagner、Christine Ackermann、Alexander Fehre、Alexander Fehre、Christian Kirschenmann、Tim Lessmann、Jakub Pakula、Hakan Sakarya、Jörg Schmitt、Moritz Köhler、Daniea Schröder、Julia Weigle、Frank Faßmer、Axel Knapp、Yuan Peng、Michael Bertram、Elena Nuthmann-Maysyuk、Klaus-Dieter Nuthmann

参与设计：Pfarré、München、Einkauf & Logistik、Stuttgart、Theapro Planungsgesellschaft、München、DS-Plan、Stuttgart

设计公司：伊波利托·福来茨集团
项目地点：乌兹别克斯坦
建筑面积：40,000平方米
摄影师：Zooey Braun、Andreas J. Focke

glazed facade. Behind illuminated facade columns of Thassos marble, a similarly weighty epochal semicircular Swarovski chandelier dominates the main foyer. Gigantic and contrastive, the two structural elements of columns and chandelier, conjunctions of various architectural traditions, introduce syntheses of Western and Eastern culture. Three marble portals with carved wooden doors lead into the vestibule. The generosity of the main foyer with a ceiling height of over 16 metres and an area of 2,500 square metres is deliberately staged and atmospherically compact. The spectacular chandelier with its nine-metre height and 23-metre length outlines the longitudinal axis of the foyer.

AUDITORIUM

Six portals of highly-polished ebony provide entry into the Auditorium. Subdivided horizontally into two sections, the gigantic hall is a place of inspiration. With a height of 48 metres and a diameter of 50 metres, it is of imposing capacity. Thanks to precise lighting technology, the bands of florescent and LED lighting that run off-set around the rings transform the space into a weightless body of light as required. The stage mechanics and technology also meet the highest of demands. Cutting-edge technology provides optimal conditions for symphonic concerts, ballet or theatre performances, as well as congresses and conferences.

VIP FOYER

The observer encounters integrated design themes everywhere, throughout the building. A striking specific example is provided by the refractions of light in the VIP Foyer. Hanging from rectangular, indirectly-lit cupolas, 18 Swarovski chandeliers dominate the atmosphere. An supersized folded wall of leaf-palladium continues the play of light. Polished marble floors deepen the

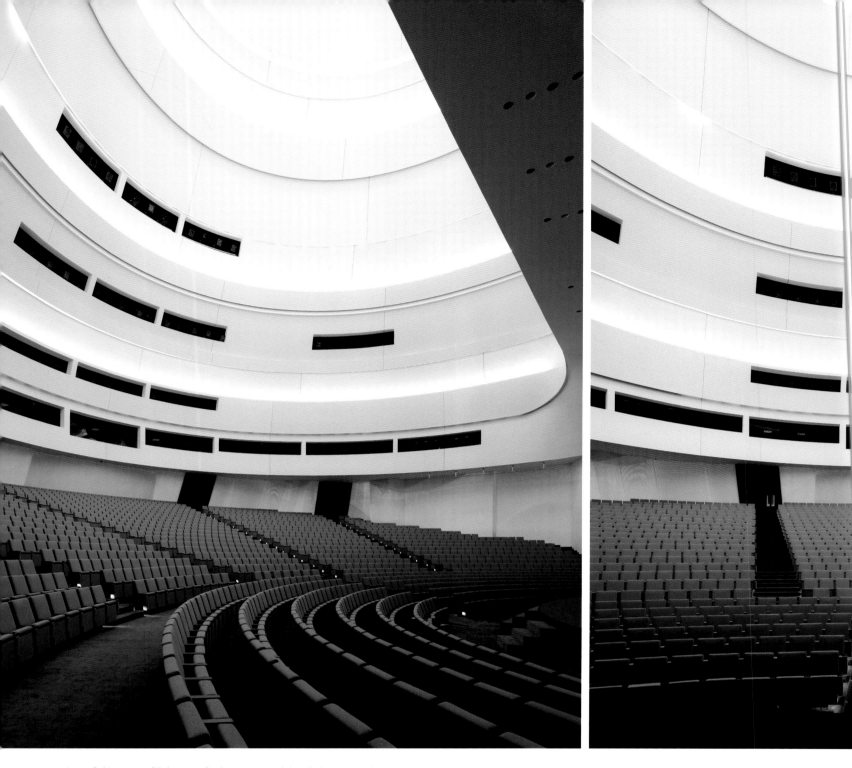

perspectives of this room of light. As a further integrated detail, the mirror elements double the spatial axes along the windows, as already seen in the foyer.

BANQUETING HALL

Aligned along the middle axis, stage and presidential area are visually connected in the Banqueting Hall. A further consolidating element is the precious nacre wall rising dramatically to the ceiling. To the left and right of the stage, optically undulating walls coated with Swarovski wallpaper strive toward the presidential area. This undulating wall structure appears as a continuation of the stage curtain and produces additional depth through the effects of reflection.

CONFERENCE ROOM

The room for international summits is dominated by an exclusive round table with a diameter of 10 metres. Linear light rings of sycamore and dark rings of ebony are inlaid in the made-to-measure walnut piece of furniture. In its presence, the table provides the representative and stoic focus for work.

RESTAURANT

The 28-metre-long buffet is placed in front of a wall of ebony. The free-flow layout, designed in stainless steel and glass, can serve up to 3,000 guests. The preparation of the dishes is carried out with energy-efficient kitchen technology and gastronomic equipment of the premium segment. The representative hospitality technology with several cold and warm kitchens, in-house confectionery, butchery as well as a presidential kitchen with attached laboratory covering more than 1,700 square metres is augmented by a further 3,200 square metres of wide-ranging production and technical utility rooms to guarantee optimal support.

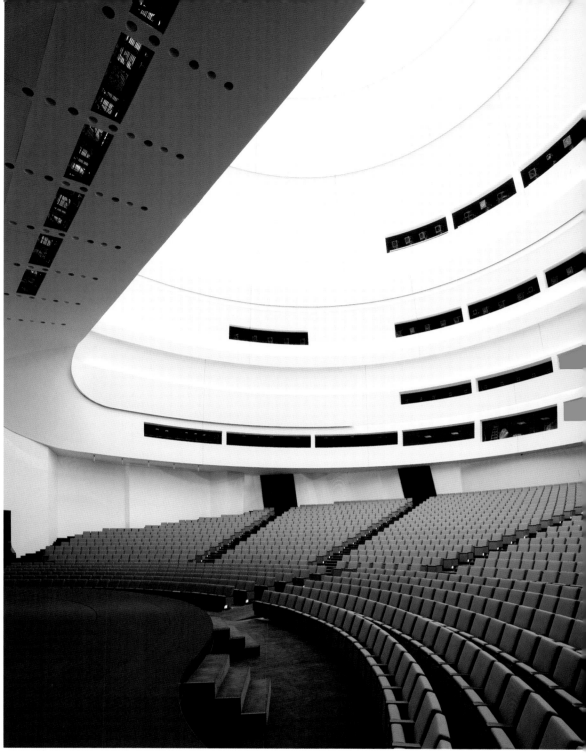

乌兹别克斯坦国际论坛宫殿矗立于塔什干市中心的阿米尔提姆尔广场(Amir Timur Square)上，是该国最重要的代表性的建筑，它被设计成举办国家活动、各种会议和其他重要文化活动、代表大会的平台。我们的任务是赋予内政一个当代的形式，同时融入了传统的乌兹别克斯坦建筑元素。其结果打造是一个由专用的材料包裹着的世界性的、长于沟通的内政。平面装饰、有机运动、水晶、贵重金属和人工灯光以及自然光的相互作用所产生的迷人景象都成为灵感的源泉。

门厅

古典的外观，此建筑准备通过被广泛运用的玻璃幕墙来展现其内在的现代性。明亮的幕墙后矗立着Thassos大理石柱，同样重要的具有划时代意义的半圆形水晶吊灯装饰着整个大厅。石柱和吊灯，两种不同的建筑元素形成巨大的反差，各种建筑构件传统的结合导致了东西方文化的融合。带雕刻木门的三个大理石入口直通大厅。高16米，占地2500平方米的大厅经过精心设计，其慷慨大气之势极其浓厚。9米高和23米长的吊灯，勾勒出了大厅的纵向轴线，一片壮观景象。

礼堂

六个高抛光黑檀木门是礼堂入口。巨大的大厅横向分为两部分，是激发灵感的地方。高48米，直径50米，容量极其巨大。由于精确的照明技术，荧光灯和LED照明灯发出的光束将其打造成光的失重空间。舞台机械和技术也达到最高水平。尖端技术为交响音乐会，芭蕾舞和戏剧表演，以及代表大会和会议创造最佳条件。

贵宾休息室

仔细观察就会发现整个建筑内集成设计的主题无处不在。一个突出的例子就是贵宾休息室中的光折射。间接照明的冲天炉，垂直悬挂的18个施华洛世奇水晶吊灯营造出整个气氛。一个超大型的折叠叶钯墙充分利用了灯光

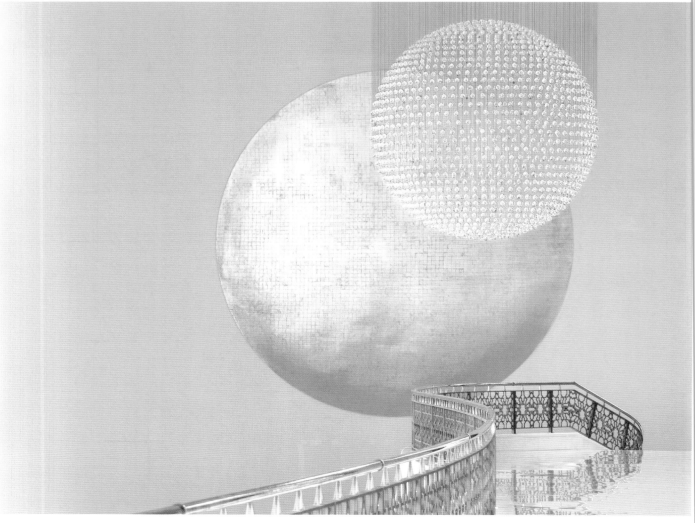

效果，抛光大理石地板搭配深房间光线的景致。为了进一步整合细节，镜面元素让空间轴沿窗户增加一倍，正如已经在大厅看到的一样。

宴会厅

沿中轴线，可以看到舞台和主席台与宴会厅是连在一起的。作为进一步巩固元素，珍贵的珍珠层壁直升天花板。用施华洛世奇壁纸装饰的波状起伏的墙壁从舞台左右两边一直伸向主席台。波状起伏的墙壁看起来就像舞台连绵的帷幕，并通过反射这种效果被再次加深。

会议室

国际峰会会议室主要放置了一个10米直径的特制圆桌。悬铃木的明线环和黑檀木的暗线环被镶嵌在精心制作的胡桃木家具中。该桌子的作用就是让与会人员专注于工作。

餐厅

28米长的自助餐台摆放在黑檀木墙前面。自由流动的布局，由不锈钢和玻璃建成的餐厅能容纳3000名客人。美味的菜肴经由节能厨房技术和高端烹饪设备制作而成。冷盘、热食、自制甜食、屠场以及带有面积超过1700平方米实验室的总统厨房，这些都是富有代表性的食宿招待技术，同时他们的功能被进一步放大，由一个3200平方米的生产和技术运作室为其提供最佳支持。

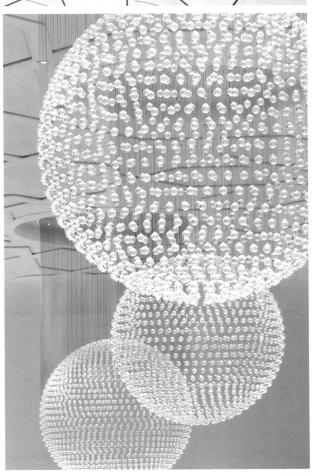

Shenyang Baoli Twelve Oak Manor Club
沈阳保利十二橡树庄园别墅会所

Shenyang Baoli Twelve Oak Manor is located at Qipan mountain, a national scenic zone, where the environment is quite gorgeous and tranquil, pictured with garden scenery in California of America. The case is the first garden project condensed with rich experience in villa developing of twenty cities of the country in last 18 years, since Baoli estate have launched in Shenyang for 8 years. It is aiming at international mansion, trying to overturn the city top-rank villa standard, so as to introduce the villa field to the world.

We intend to improve the quality of architecture and living comfort by means of utilizing exquisite design from exterior to interior with natural stone material, high-range solid wooden door and window and high-tech installations, from which we fully present royal sense and imperial temperament of north American garden. Moreover, we display an American custom and region traits of this case in a brand new way, with every space owning unique understanding of American garden style and demonstration towards leisure life conception.

设 计 师：王赟、王小峰
设计公司：广州尚逸装饰设计有限公司
建筑面积：1700平方米

095

　　沈阳保利十二橡树庄园，位于沈阳市国家级风景区棋盘山，地域环境优美、静谧，极具美国加州田园景观。该项目是保利地产进入沈阳8年后，凝聚全国十八年、二十城的顶级别墅开发经验，实力打造的首个庄园项目。以国际豪宅视野，颠覆城市顶端豪宅标准，引领豪宅领域走向世界。

　　保利十二橡树庄园，通过由外及内的精雕细琢和对天然石材、高档实木门窗、尖端科技的配套等，致力提升建筑品质及居住舒适度，全景体现北美庄园的贵族血统和王者风范。以全新的方式展示该项目的美国风情与地域性，每一个空间都显现对美式庄园独到风格的理解和诠释休闲生活概念。

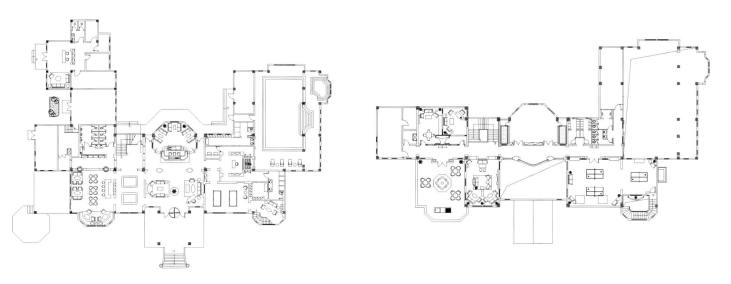

Farglory New City

远雄新都

Regarding the appearance, we establish this business building into a privileged existence in the boundless earth of local area. In the daytime, the architecture stands straight like a landmark, imposing and spectacular, which resembles an art museum that people are eager to explore. When the night approaches, the top tower sends out light illumination, besides glasses of the building with net print are reflecting brighter light beams, just as a piece of diamond with inner shine.

Along the parking lot, people can walk on the gradual high ramp, which is cleverly used by us to disclose the prelude of the pilgrimage trip. The reception center looks like rectangular shape from outside, but in fact we establish a round architecture as a tall cathedral. The real meaning of English word "architecture" just represents the significance formed by "arc" and "tec", which means "the art technique that created by arc shapes". Thus "building inside building" is exact as the soul of real object, attracting customers to visit, compliment and explore.

设 计 师：黄书恒
参与设计：欧阳毅、陈新强、蔡明宪、胡春惠
　　　　　胡春梅
设计公司：玄武设计 Sherwood Design
项目地点：中国台湾
建筑面积：3075.6平方米
主要建材：塑铝板、矿纤水泥板、大理石、型网

We intentionally lower the entrance highness. Once entering the dome building with 17 meters high, visitors will be impressed with grand and broad prospect. In addition, we specially cut the dome board by unique proportion, making the whole space more tall and solemn. The visitor seems to come into a theatre with a dome. There are black stairs embodying with stage effect, which seems to wait for the arrival of protagonist. All of sudden, the atmosphere is condensing that a pious and devout feeling spontaneously rises out of nowhere.

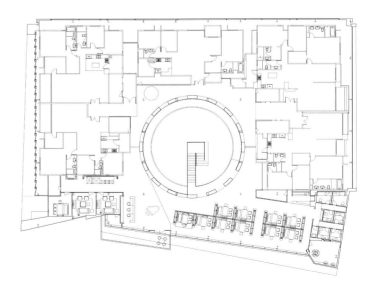

在外观上，设计者将此商业建筑，塑造成为在当地一望无际的视野、当之无愧的优越建筑——白天，耸立如地标，雄伟壮观，正如让人渴望一窥堂奥的美术馆；入夜，其上光塔散发光源，建筑本身所使用的网印玻璃更透出清亮的光线，正如暖暖含光的金钢石。

顺着停车场，沿着缓步高升的坡道拾级而上，设计者巧妙运用细节揭开这程朝圣之旅的序幕。外观看似长方体的接待中心，却在其中打造出挑高如圣堂的圆形建筑。"建筑"architecture一词的英文意义，正代表着arc.（弧）与tec.（技术）所形成的"由弧形拱组成的工艺技术"。而这"建筑中的建筑"，正如实体中的灵魂，吸引着参访者绕行、赞叹、探索。

入口高度刻意压低，让访客一进入挑高达17米的圆顶建筑，顿觉堂皇开阔。此外，设计者更特意将打造圆顶的板块以特殊比例切割，使整个挑高空间更显高耸庄严。访客彷佛来到一座圆顶剧场，剧场中更有一方兼具舞台效果的黑色阶梯，似乎随时等待着主角现身降临。气氛在此凝结如水，一种莫名的虔诚圣洁之感油然而生。

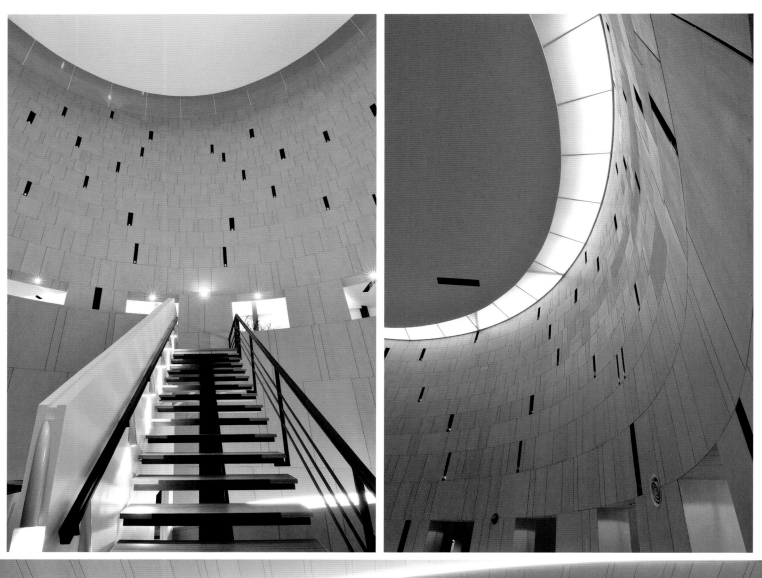

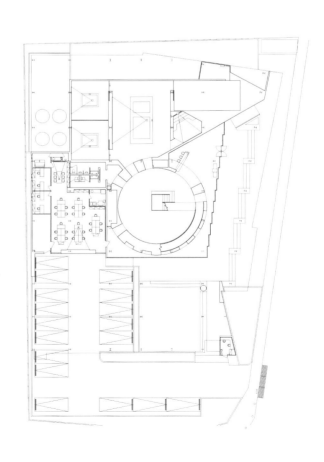

设 计 师：沈中怡
参与设计：杨佩珩、江易书
设计公司：中怡设计公司
项目地点：中国台湾
建筑面积：一楼:567.6平方米 二楼:683.1平方米
主要建材：钢构、玻璃、南方松、铁件、镜面不锈钢、黑镜、石材

Qingjing Reception Hall

早安清境接待会馆

On one side of the base itself is facing a road along the creek, the other side is facing the future plans (when planned, not yet opened), therefore there isn't difference between front and back. The designer designs the whole building appearance as egg-shaped, creates a continuity surrounded visual effect, and adopts "double" concept. Innermost layer is the quantity body stack of the functional development, plus a layer of egg type real wood grille on surface and echoes with the natural environment factor of base. He uses grille as the surface material, which makes "light" and "shadow" become part of the spatial elements naturally.

The whole interior space uses simple but elegant wood texture echoing with natural outdoor scene and white tone space, which is both background and light shadow dancing stage. Overall presents as leisure comfortable tone of a gallery, the indoor dynamic lines are deliberately arranged, the best view is configurated to the negotiate area, which allows visitors' interaction to link with outdoors scene.

　　由于基地本身一侧面临沿溪小路、一侧面临未来的计划道路(规划时，还没有开通)，因此并没有正背面的区分。设计师将整栋建筑外观设计为蛋形，创造一种连续性环绕的视觉效果。同时采用"双层"的概念，里层空间由机能发展的量体堆栈而成，表层则加上一层蛋形实木格栅，呼应基地所处的自然环境因素。运用格栅作为表层的素材，很自然地让"光线"及"阴影"成为空间元素的一部分。

　　整个室内空间以素雅的木料质感呼应自然的室外景致，并与白色基调空间相呼应，既是背景也是光影舞动的舞台。整体呈现如艺廊般的休闲舒适基调，室内动线经由刻意安排，最好的视野则配置给洽谈区，让参观者的互动能够与户外景致连结。

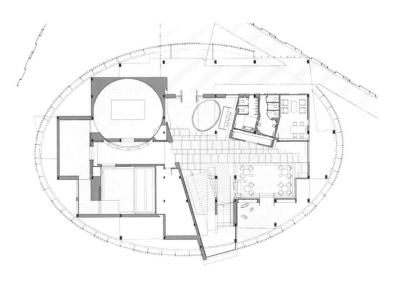

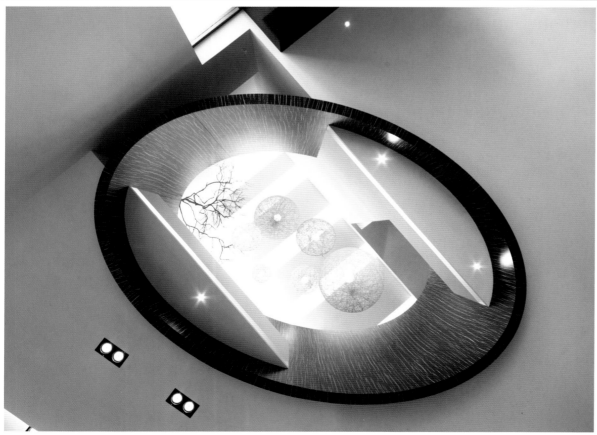

123

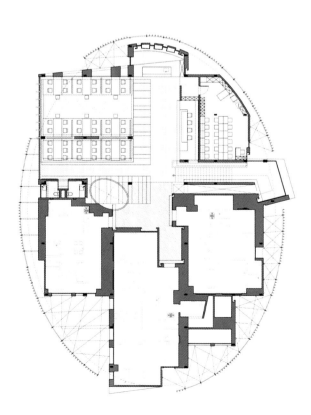

The Ideal Club
如意会所

The dome looks natural and smooth, like floating clouds and flowing water.

This case is located in the Pangu Plaza (also known as the Seven Star Morgan Plaza) which is in the west of Beijing National Stadium (also known as Bird's Nest), with an area of 500 square meters, making the club form as the main spindle of the design. The overall layout of the internal space is a central hall connecting with three exclusive VIP rooms which are independent of each other, highlighting the personal temperament. The "ideal" is the main spirit of the design of this case, the integration of clouds, Ganoderma lucidum, ideal. These convoluted coiled specious flowers, leaves, branches and tendrils do have an air of clouds, luxury but non-publicity, restrained but full of profound background, fully expressing the club's temperament, character and create the perfect atmosphere in its space, and also encouraging an artistic life of indulgence and pleasure. Entering into the corridor, you can find architectural lintels made of cobwebbing rose gold stainless steel, fine leather surface, goldleaf dome, and Ariston stone floor, which mean that the guests here are rich or important. A space with accommodation sky and silver roof can be found in the hall, which is a magnanimous scene and imposing. The design of vault adopts LED fiber optic lights to create an effect of floating clouds and

设 计 师：王俊钦、彭晴
参与设计：赵文静、曹永辉
设计公司：睿智匯设计公司
项目地点：中国北京
建筑面积：500平方米
主要建材：镜面、拉丝玫瑰金不锈钢、牛皮、银箔、金箔、壁纸、茶镜、明镜、橡木饰面板、西班牙米黄石材、雅士白石材

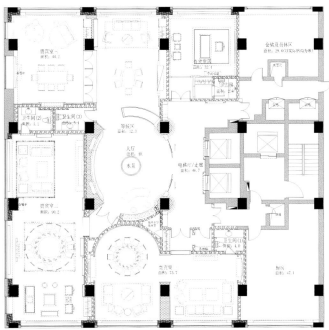

flowing water, and the lines of floating clouds and flowing water are used to sketch out a picture of low-key luxury. And the leather walls are decorated by metal buckles with carved imagery flowers, adding a sense of steadiness. There are crystal chandeliers, red walls, priceless antiques, wall fireplace, top European furniture in waiting area at the hall. When going into it, you can feel as if you are joining in a salon of European aristocracy through a time tunnel.

The co-existence of extravagance and simplicity

The guests can enjoy a different style in three VIP rooms, which is different from other clubs. At the beginning of the design, designers plan the function of the place from the users' point, supplying sophisticated and comprehensive services to meet the actual needs. And each room, which is distinctive and with excellent privacy, is definitely a good social gatherings place for the celebrities and VIPs. The actual luxury comes from no jade and gold, but from art and taste, by this way, you can enjoy yourself as if you are in a palace. Chinese VIP Room is the center of this case, which mainly adopts Chinese design style. The space is divided into reception area and dining area. Use the form of open space, the overall design is not a traditional Chinese style, but a simple and stable style. Interior sprung roof is made up of the stack and rotation of silver foils, with the ornate crystal chandeliers, the space is decorated to be brilliant. The wall is decorated by the ebony and yellow coloured glass which show a sence of steadiness and reality. And the Chinese Baibaoge is added in it by the contrast of metal materials, which bring out the crucial point. In order to increase Chinese background and the visual impact, post-modern techniques are taken to have the simple but luxury European-style furniture and crystal lamps combined with Chinese furniture in modern style, which make a splendid space. The use of a lot of leather, silk, saddle stitched furs, and printed fabrics will let you have a feeling of falling into a luxury ocean. The luxury of the furniture here can be called "the luxury on the hair," every detail is showed so perfect.

穹顶，行云流水之势

本案地处北京鸟巢西侧之盘古大观七星摩根广场内，面积为500平方米，以会所形式为设计主轴，内部空间以中心大厅连接三大专属贵宾室，贵宾室彼此间独立而至，彰显私人气质。以"如意"为此案设计主精神，"祥云、灵芝、如意"，它那旋绕盘曲的似是而非的花叶枝蔓确得祥云之神气。奢华而不张扬，内敛而彰显丰厚底蕴，把会所的气质内涵、性格的彰显、氛围的营造完美展现于空间脉络中，也鼓动着放纵而享乐的艺术生活之感。初踏走廊过道，拉丝玫瑰金不锈钢的建筑造型门楣、细腻牛皮式曲面、金箔式穹顶，雅士白石材地面，无一不默默地彰显出入此间的客人非富即贵。进入大厅，圆融之天穹银顶空间，大气的场面，震慑全场。天穹之处以LED光纤灯营造出行云流水之效果，并以"如意"祥云流水之线条勾勒出低调奢华，墙面以金属扣用意象雕花呈现于牛皮墙面，更增添空间之稳重。大厅处的等候区，水晶吊灯、红酒幕墙、价值连城的古董、墙面壁炉、欧式顶级家具，步入其间，仿佛穿越时光的隧道，进入了欧洲的贵族沙龙。

奢极至简

三大贵宾室更以不同设计风格呈现给享用者，有别于一般的会所设计。设计之初就从使用者角度策划空间功能，以更精致且完善的服务切合实际需求，各不同包厢个性鲜明，私密性极好，是名流贵宾的社交聚会之处，真正的奢华决非金碧辉煌的堆金砌玉，艺术与品味的相映生辉，才能成就殿堂级的奢享。中式贵宾室为本案中心，此空间以中式设计为主，空间分为会客区及用餐区。空间为开放形式呈现，整体设计并非以传统中式表现，而以简约并稳重的方式展现。室内吊顶用银箔面叠加并旋转，配合华丽的水晶吊灯，把空间装饰得流光溢彩。墙面以黑檀木与茶镜虚实表现稳重质感，并以画龙点睛之形式，将中式之百宝阁以金属材质的反差点缀其间。为增加中式底蕴及视觉冲击，采用后现代主义手法，将简约式奢华的欧式家具及水晶灯与现代手法中式家具相互结合，围合出一个气派的空间。大量皮革、绸缎、马鞍缝法的皮毛、印花织物的运用，让人一进入其中就仿佛跌进奢华的海洋。这里家具的奢华可以被称为"头发丝上的奢华"，每个细节展现得是那么尽善尽美。

Shanghai Huawu Club

上海华屋馆

The case is located in Shanghai villa agency and also is the first national company defined by the top estate honorable service, whose main target is to turn a new leaf for real estate by focusing the business on top VIPs. We embark on the design from defining the space with material, furniture, lights, space figure and music combination, so we break the agency's stereotype system rule. And a new outlook that estate information is merging with public media, which accomplishes a multiple variable space.

Following the generatrix, we come into the model region, a displaying room in exquisite arc shape. The white drapery for lampshade spreads from the ceiling. The round shape both in the ceiling and on the floor are echoing with each other, which is cleverly sending out a message of being complete. The interior room has three floors, each of which is actually not very high. One of our designing point is to ensure that every visitor can feel boundless free atmosphere without any pressure even in the limited space. As to the private conference room, also the area where the buyers would stay for the longest time, for what it is worth, we

设 计 师：黄鹏霖、黄怀德
设计公司：台北基础设计中心
项目地点：中国上海
建筑面积：2640平方米
主要建材：黑云石大理石、金锋石大理石、实木喷漆、金属喷漆、黑镜

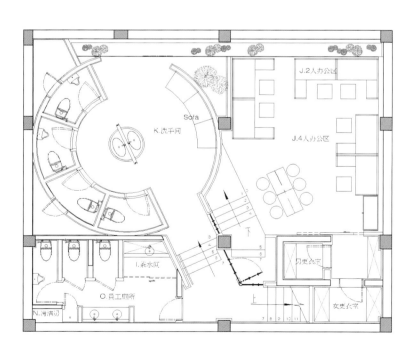

release the most spacious vision effect for every visitor. The window has a perfect vertical and horizontal stretch without affecting the structure of architecture, so that it can make the most of daylights inside. With the abundant daylights and nice floor weaving out a warm tone, and the green outside swaying cutely, the visitors can really find the cozy feeling that only exists at home, when they are negotiating.

We resort to mixing match approach to make the space burst out generous non-cold but warm vision, which is definitely not going with the flow. The marble luxury texture on the wall of conference room nicely neutralizes the floor temperature. The cutting dark mirror on the ceiling makes the inside building appear gorgeously tall. Switching to an atmosphere of leisurely peaceful nature music, people can be intoxicated in the aesthetic life style.

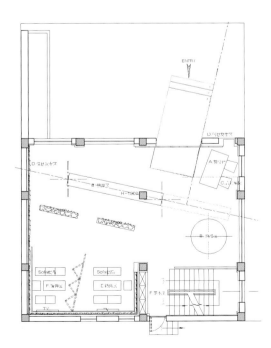

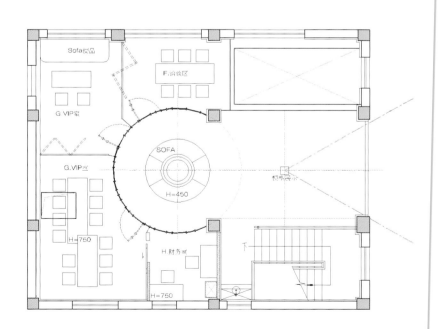

此案位于上海豪宅区的房产中介公司，也是全国首家定位于高端不动产尊贵服务，针对顶级VIP为经营主轴的新地产面貌，从空间开始定位，借由材料、家具、灯光、空间线条以至音乐结合动线等，打破传统中介公司的制式规格，以崭新的面貌出现，打造结合多媒体的房地产信息，成就复合式的多元空间。

沿着动线来到模型区，精致的圆弧形展示空间，自天花板延伸出白色帷幕罩着灯源，天地双圆呼应，巧妙地传递出圆满的凝聚意象。三层楼的室内空间，其实每层楼的楼高并不高，在有限的楼高中，如何让参观者能感受到最自在无压的氛围，是设计师形塑此案的重点之一。随动线来到较为私密的洽谈区，也是购屋者会停留最久的区域，因此设计师将空间释放出最宽敞的视觉效果，在不影响建筑结构的前提下，窗户的垂直与水平皆延伸极致，尽可能将光线揽入室内，充足的采光与温润的木地板交织出温暖的调性，窗外绿意摇曳生姿，让参观者在洽谈的同时，能获得如在自家中的舒适感。

以混搭的技巧让空间释放大气却不冰冷、暖馨却不流于俗的空间视感。洽谈区墙面以大理石的奢华质感中和木地板温度，延伸至天花板，取而代之的切割黑镜让楼层高度显得挑高富丽，在转为平静悠缓的自然音乐下，细细地品味优雅的美学生活。

Queen's Cube
女王的魔方

The showroom of the sales department of Queen's Cube is located in a commercial building in the center of the city. The designer shows Queen's Cube idea in a limited area. The inspiration of the designer comes from the sight around the city, starting with the community environment. This is designed to give the guests an experience journey and then to establish an alternative attitude towards life. Designer take the idea of "Life is struck by a new wave" as a creative concept. "A New Wave" stands for a new lifestyle in this particular region, which is full of a new force. He uses symbolic meanings to express the idea in the exhibition hall, convey a living culture of settlement patterns integrating into the urban.

Entering into the showroom of the sales department, you can see the arc element. Carrying on the concept of the wave, the designer use the buildings with light brown stripes to surround the exhibition hall. Irregular pattern showing the innervation of wave and urban concept map as the background shows a picture of the wave surrounding the city, meaning the whole place linking together.

In order to express the regional character of this place, the designer adopts an imagery map to replace the stardard area map in the design of area display, using pattern to indicate the precious cultural heritage in Wan Chai, for example, trams, temples, ferry, etc., by this way, the guests can get a deep impression of the local

设 计 师：何宗宪
设计公司：何宗宪设计事务所
项目地点：中国香港

history and development status.

The design of the wave can be found from the demonstration unit showing the actual living space to business negotiation area. Gorgeous crystal strings, which means the spray from the waves beat and the waves caused by the sales experience, surround the display area of the main model. The business negotiation area also adopts glass pattern of spray. The guests and the land agents can discuss some details in the surrounding of waves. The design of the whole building keep to the point of the theme of the wave. In addition, the overall atmosphere of the showroom and the lighting processing make full use of the night scene in the urban to show the bright side of the city, which allow the guests to feel the city's vitality and experience a new lifestyle.

Queen's Cube的售楼部展示厅设于市中心的一幢商业大厦内,设计师在面积有限的展厅内把Queen's Cube的理念展现出来。该住宅项目周边的城市景象给予设计师灵感,以社区环境为设计出发点,目的是为客人带来一个体验的旅程,以及建立起另一种生活态度。设计师以"Life is struck by a new wave"("生命中的新浪潮")为创作概念,"新浪潮"代表在这特别的地区有新的生活模式,当中充满一股新的力量。他们用象征式的手法把构思呈现在展示厅内,表达居住模式融入都市本身的生活文化。

走进售楼部展示厅内,可以见到弧线元素。承接"浪潮"的构想,设计师以浅棕色的曲线条子建筑包围整个展示厅,不规则的设计表达出波浪的动感,以都市概念图作为背景,比喻浪潮包围整个城市,意思是把整个地方连结在一起。

为了表达该地点的区域个性,设计师在项目的地区展示方面以意像地图取代标准的地区地图,利用图案标志着湾仔区内珍贵的文化遗产,例如电车、庙宇、渡轮,让客人对区内的历史及发展留下深刻印象。

由展现实际生活空间的示范单位到商谈生意的洽谈区连贯了整个浪潮设计,华丽的水晶串代表波浪拍起的水花,包围着主体模型展示区,也代表整个售楼体验所引起的浪花。而洽谈区也利用玻璃浪花图像,让客人和地产商在浪花的包围下商讨细节,生动地紧扣以浪潮为题的创作蓝本。此外,展厅的整体气氛和灯光处理是借着大都会的夜晚表现出城市璀璨的一面,同时让客人感受城市的生命力,体验一种新的生活模式。

Jinan Other Mountains Club
济南他山会所

The phase of "other mountains" is pregnant with ancient meaning. However, "other mountains" sale center is trying to pursue a suitable environment for humanities. Although we adopt all sharp modern materials, with classic match of black and white, to build a fashionable exterior appearance, the interior is spreading not only modern Chinese style but also an expression of humanity emotions and art significance.

Once opening the door, people can see the theme wall of mountain picture, with pure white color tone and smoothly fluent figure, and all those "little hills" is reflected by the mirror like flowing water that never ends. They travel from this piece of wall to another, charming like flowing water. The little fountain in front of the theme wall, green bamboo tube in the big container and the soft water sound are performing nature music. The tough pole and ceiling can reflect graceful figure of "mountains", naturally forming a beautiful landscape painting when it comes to an echoing moment.

Every giant and firm metallic pole symbolizes the changeable bamboo. And with the sense of bamboo, the atmosphere is getting more nature setting. The designed huge paper drop-light is based on mountain streams in Chinese ink and the cover of black wire lampshade makes it more elegant and piercing. In the middle night,

设 计 师：戴勇
设计公司：戴勇室内设计师事务所
项目面积：500平方米

the light is still on and the quiet water stream starts flowing to every peaceful corner, making the lonely night begin evaporate liveness.

The light is an object, which can be designed in many different shapes. Illuminating on the original static objects, the light can evoke vigor from deep inside also can pick up a gracious painting roll. The layout is not the only part that we elaborately design, we also pay much attention to the color adoption. The special pink painting glass on the table is extremely enchanting under the pure lights. Popping into people's vision, it can bring happiness and warmth, fashion and dynamic.

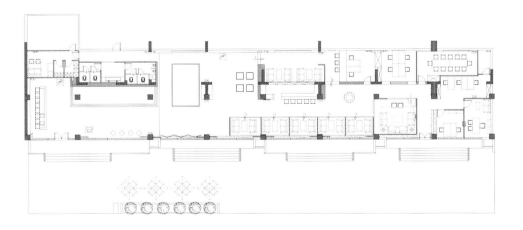

"他山"深蕴古意之词,此处"他山"销售中心意在追寻人文居住的舒适雅境。虽用酷睿的现代感材质、经典的黑白搭配,构筑时尚外观,然而室内延续的不仅是现代新中式的韵味,更融入一份人文情感及艺术表现。

开门即见山体造型背景墙,纯美的柔白色调,线条自由休酣,那一座座"小山峰",在镜面的承接下绵延不断,绕过一道墙,转到另一面墙,仿似水般柔顺妩媚。主题墙前的小小水池,大缸中的青翠竹筒,缓缓流动的轻轻水声,共同演奏自然的乐章。刚硬的墙柱与天花倒映出"山峰"的柔美曲线,灵动处自成一幅山水美画。

一根根高大坚挺的金属圆柱是变幻后的竹子,带着竹的气息,进一步烘托自然的氛围。设计定做的大型纸吊灯以水墨山涧溪流为题材,黑丝灯罩的外套让它显得幽雅深邃,夜幕中灯光依次照亮,静默的溪流开始缓缓流动,流向每个寂静的角落,让寂寥的夜开始散发生气。

光是物体,可以设计为多种表现形态,照射在原本平静的物体上,能勾起深底里的活力,勾勒出唯美的画卷。设计师精心设计的不仅是造型还有色彩,桌面上特别添加的桃红色油漆玻璃,纯净的光线让它倍加姣美妖娆,跳跃在视线里,带来温暖与欢愉,时尚与动感。

Grand Resort Club Bad Ragaz
巴德拉格斯酒店会所

设 计 师：Joseph Smolenicky
摄 影 师：Roland Bernath、Walter Mair, and Smolenicky&Partner
设计公司：瑞士圣加仑州戈绍Blumer–Lehmann AG公司
项目地点：瑞士巴德拉格斯
建筑面积：90,000平方米

LANDSCAPE SITUATION

The town-planning character of the resort is dominated by large representative buildings set in an expansive park landscape. To this extent, the resort clearly distinguishes itself from the identity of the village of Bad Ragaz. During the belle époque, this principle of building monumental hotels in close vicinity to a village was successfully applied to a number of locations in the Swiss Alps. The most important examples are Interlaken, St. Moritz and Gstaad. In Bad Ragaz, two cul-de-sacs fork off from the main road that runs through the golf

course connecting Bad Ragaz and Maienfeld. In the new project the thermal baths were deliberately located on the cul-de-sac accessing the resort's public facilities, such as the new conference centre in the renovated spa spring hall, the casino and the golf club house. The second cul-de-sac running along the park has been kept free to provide access to the three grand hotels, and emits a more private and calmer atmosphere.

FORM AND EXTERIOR SPACE

Instead of being freestanding, the form of the building volume emerges from the enclosing of exterior spaces. In the area of the open-air baths, for instance, the volume of the building is stepped back and opens out the sunbathing lawn to the wooded slopes of the mountain ridge. The view extends past the existing buildings, screened by newly planted groups of trees. The guests experience a park landscape that melts into woods and mountain slopes.

The predominant landscaped, park-like atmosphere remains intact despite the compact manner of building. Thus the resort remains characterized by its park. The main entrance to the thermal baths, the spa spring hall, is set on the visual axis of the cul-de-sac in order, from the main road, to mark its presence in the depth of the site as a public facility.

CRITERIA OF THE BUILDING

The Tamina thermal baths is explicitly conceived as a part of the grand-hotel culture. The cultural and aesthetic identity of the project seeks an affinity to both Swiss tradition and the grand hotels of the Baltic coast, such as Heiligendamm.

For this reason the building volume has a monumental character, in order to stand out as an institution equal to the other buildings in the resort. Simultaneously the thermal baths are intended to relativize the almost "urban" stonework character of the spa spring hall. This explains the snow-white woodwork of the thermal baths, lending it the pavilion-like character of the architecture of a historical holiday resort.

This strategy of using an explicit resort architecture is underscored in the building's formally fanciful oval windows. Seen from the inside, the windows have the effect of over-dimensional picture frames. Oval picture frames were widespread in the Victorian era for landscape scenes, whereby the intention in the current

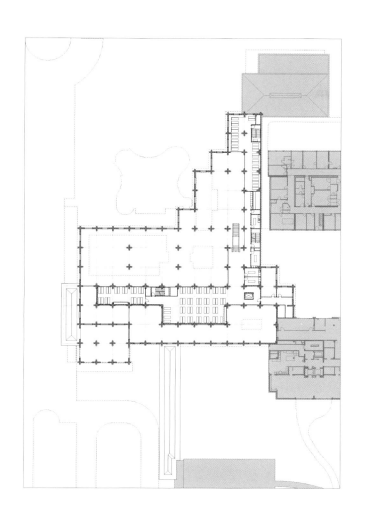

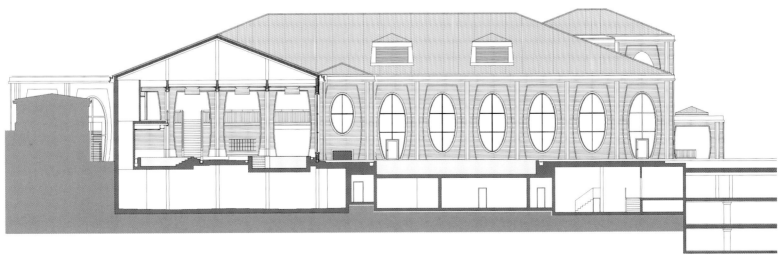

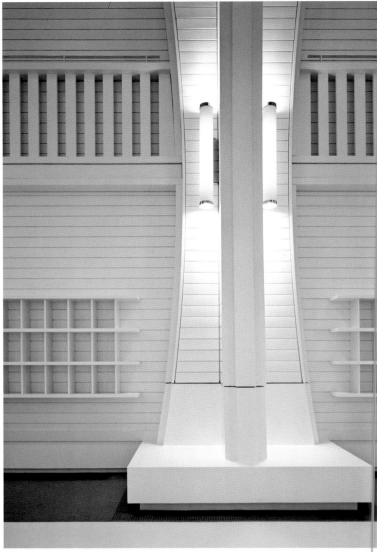

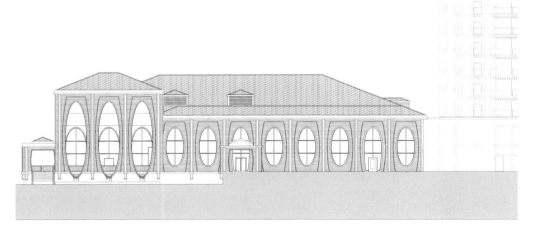

project is to give specific expression to the view over the relatively neutral landscape by means of the gesture of the frame.

INTERIOR SPACE AS A "FOREST"

Metaphorically the creation of the interior spaces of the project has an analogy in cutting clearings in the pattern of a forest by felling individual trees. This is the reverse of the common design process. The exterior spaces are similarly created by "felling" supports on the periphery of the building volume. Structurally the building can be more or less seen as a forest, created out of columns instead of trees – a total of 115 supports using the timber of 2,200 fir-trees (this amount of wood is regenerated in Switzerland in two-and-a-half hours).

EXPRESSION AND MATERIAL

Materially the project possesses the same appearance internally and externally. The snow-white timber battens are carried over internally as wall surfacing. In this sense there is no actual interior architecture to the building, but instead only a whole architecture of the building.

The timber structure of the building is not merely determined by the criteria of the span of the supports. Instead of a focus on the engineering of the function of the supports and the reinforcement of a construction, the structure concentrates far more on spatial phenomena, creating a beauty and a ceremonial atmosphere. Bathing is celebrated as a cultivated activity.

建筑物场地

本案商业区规划的特点是将各楼建成大型的有代表性的建筑物,嵌入于昂贵公园景观处。从这个意义上说,该商业区已清晰地从巴德拉格斯的乡村中分离出来。在一战前的美好时期,在村庄附近建立有纪念意义的酒店的观念,在阿尔卑斯山上的许多地区得到了非常成功的体现。其中最重要的例子是Interlakn酒店、St. Moritz酒店和Gstaad酒店。在巴德拉格斯,从大路上分支出来的两条死胡同穿过连接巴德拉格斯和迈恩费尔德的高尔夫球场。新建的温泉刻意选在死胡同里,由此可进入酒店的公共场所,比如整修过的大厅里的新会议中心、赌场和高尔夫俱乐部。第二个胡同沿着公园,能够免费通往三大酒店,营造着一种更隐秘和宁静的气氛。

建筑物造型和外部空间

本案不是独立式的立体建筑,而是被外部空间包围着。比如户外洗浴区,建筑物会稍稍靠后,同时日光浴草坪向山脊上树木丛生的斜坡伸展开去。透过建筑物看另一侧,视线将被一排排刚刚种植的树遮挡。在此顾客能够体验树木与山坡融为一体的公园景观。

充满公园般气氛的景观依然保持着其完整性,尽管受到施工的影响。本案也仍然保持着它本身的特色,其主入口可以通往温泉中心。该入口设在易见的死胡同轴线上,以便从大路上也能够看见胡同深处这个公共景观。

建筑物标准

塔米纳温泉浴场鲜明的大酒店文化及其对美的认可,与瑞士传统文化和波罗地海岸的大酒店,如Heiligendamm酒店文化产生了共鸣。

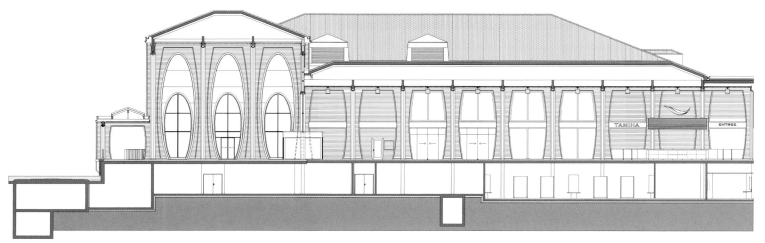

　　本案在建筑上有个很大的特点，即作为一个会所但与酒店内其他建筑物一样突出。同时，温泉浴场试图与SPA大厅中几乎所有散发城市气息的石雕工艺品一比高下。浴场内雪白的木制品，给整个酒店增添一种古代假日酒店富有的华美。

　　酒店采用鲜明的建筑风格，其效果比不上本案采用的新颖奇特的椭圆形窗户所产生的效果。从里往外看，椭圆窗户会产生一种跨时空的图片框架效果。在维多利亚时代，椭圆形图片边框作为一道景观广泛应用。本案使用椭圆图片边框是为了运用边框来展示一种相对黯淡的特别的视觉景观。

　　类似森林结构的内部空间

　　隐喻地说，本案内部空间的营造过程和依森林的样式砍倒单独树木从而切割出空地的过程类似。这是对普通设计过程的一种颠覆。外部空间的营造同样类似于通过"砍伐"建筑物外围的支柱。从结构上看，本案几乎可以被看做为一片森林，只是用圆柱代替了树木，共用了115根支柱砍伐了2200棵冷杉树（瑞士在2.5小时之内就能够重新长出2200棵冷杉）。

　　空间的表达与材料的选择

　　建筑物内外都选用了同样的材料，雪白色的木板平铺于内墙表面。从某种意义上说，该建筑并没有真正的内部结构，而只有一个唯一的整体结构。

　　本案中大梁结构并不仅取决于支柱间的间距标准。相对于支柱的工程结构及巩固，本案大梁结构更专注于三维空间的构造，营造出了美丽但不失庄重的气氛，从而把洗浴变成一种文明的庆祝活动。

Jinan Future City Club

济南未来城会所

The future is abstract, the future is relative, the future is fantasy, the future is full of energy...

The future city's positioning is young consumers, the domination of future world. In architectural appearance, box body first presents in sight, the "X" type of metal frame lets a person associate airport's terminal building, stylist wants to metaphor this building going through future terminal.

In the design language, slash and irregular body can bring a person uncertain, energetic and powerful feeling, but this exciting uneasy modelling cannot too propriety. As Chinese stress harmony to foreigners, the whole building doesn't make public, but fuse with surrounding environment, but interior building space is rich in sense of unrest, uncertain feeling and very powerful.

There is a great crystal modelling (fold face irregular forms are made of grey mirrors) at the entrance, large dimension, with colorful rays, like energy divergent, attracting the sight, steps to stop. In Chinese geamancy, crystal can change building internal geamancy, containing uncertain enormous energy, and young people's biggest characteristic of variability, they are future of society, hold enormous, uncertain energy in their bodies. Crystal modelling attracts young crowd, makes distance closer to them, casually, those troubled lines start to

设 计 师：戴勇
设计公司：戴勇室内设计师事务所
项目面积：169平方米

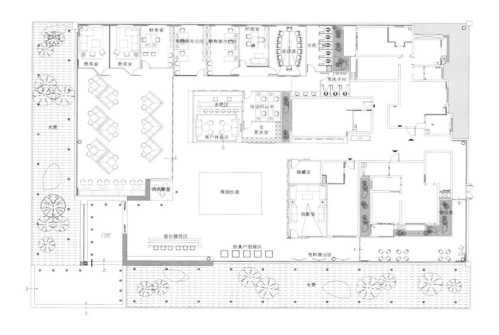

impact visual organs from various perspectives.

Lights on the ceiling presents irregular divergent shape, deep and light color alternating straight lines design carpet outspreads at the same direction, strengthens depth of the small space, the adorns interval mural on the black and white metope, abstracting is also a kind of thinking divergent. Water bar area continues slash, irregular form energy, the processing is lively and rich in rhythm, the clever apply of many grey mirrors and lighting further strengthens theme concepts, and shows containing high-tech boundless energy, a budding inrush current.

未来是抽象的，未来是相对的，未来是充满幻想的，未来是富盈着能量的

未来城的定位是年轻的消费者，未来世界的主宰者。在建筑外观上，方盒体第一时间呈现在眼前，"X"形的金属构架让人联想到机场的航站建筑，设计师想把这个建筑隐喻成通往未来的航站。

在设计语言中，斜线与不规则体块可以带给人不确定的、有活力及有冲击力的感受，但这种让人激动不安的造型不能太锋芒毕露。正如中国人对外讲求圆融，整个建筑并不张扬，而是和周围环境很融合，但建筑的内部空间是富有动荡不安的感觉，很有冲击力。

入口正面映入眼帘的就是巨大的水晶造型，灰镜做成的多折面不规则形体，体量很大，泛散开缤纷的射线，仿似能量的发散，吸引着视线、脚步的停驻。在中国风水里，水晶可以改变建筑内部的风水，蕴藏着不确定的巨大能量，而年轻人群的最大特点正是可变性，他们是社会的未来，体内蕴藏着巨大的、不确定的能量。水晶造型吸引着年轻的人群，拉近与他们的距离，不经意间那些不安分的线条从各个角度开始冲击视觉器官。

天花上的灯具呈不规则发散状，深浅色交替的直线条图案地毯朝着同一个方向延伸，加强了小空间里的纵深感，点缀在墙面的黑白间隔壁画，抽象中也是一种思维的发散。水吧区域延续斜线、不规则形体的活力，处理明快富有节奏感，大量灰镜与灯光的巧妙运用进一步强化主题概念，并昭示蕴涵一种高科技无穷的能量，一种蓄势待发的涌流。

Binge Billiards

宾格台球

The cutting method of the geometry is used in the design of the reception in the entrance and the cabinets for placing the cues in both sides. The contrast of the processed PVC pipe for the background is also adopted in this case, then the symbol of the case is embedded in it, outlining a modern geometry. The waiting area for leisure is made up of ramp-formed diamond plate, which extends the space and plays a role of partition. It not only meets the needs of the space, but also enhance the interaction between light and shadow within and outside the box. Next, the private teaching area is made by two surrounded bars available for recreational use. Although it is called "private teaching", it actually does not resist you far away from a thousand miles. The wall on the left aisle follows the uneven shape of the reception (made of wood), with display cabinets scattering among it.

The design of the masculine linear can be found in the entire space. The use of light and shadow make it a modern fashionable entertainment billiards' club. The entire entertainment space shows a beauty of a strong contrast. Even if you are not a ball friend, you will inevitably want to have a try.

设 计 师：高雄
设计公司：道和设计机构
项目地点：中国福建
建筑面积：800平方米
主要建材：黑钛、水曲柳、黑镜、木质花格、
　　　　　木纹砖、蒙古黑火烧拉槽板、墙纸、
　　　　　PVC管、地毯、绿可木

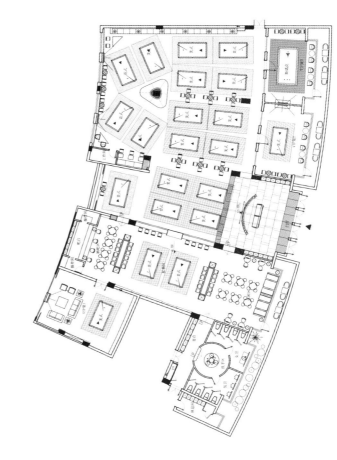

入口的前台以及两侧的寄杆柜大胆地将几何的切割法运用其中，背景的PVC管烤漆处理后产生反差。再将本案的标志嵌入其中，勾勒现代几何形体。经过金刚板斜拼组成墙面的休闲等候区，我们发现了空间的延伸，并且起到隔断的作用。不仅满足空间需求，也增强了包厢内外的光影互动。接下来，私教区是由两组可供休闲使用的吧台合围而成，虽称之为"私教"却并不拒人千里，左边过道的墙面沿用了前台的凹凸切割面造型，将展示柜分散在其中。

设计线形和块面上以阳刚硬朗为引线，贯穿了整个空间。用光影的诉说，将格调定位为现代时尚的娱乐台球会所。整套娱乐场所呈现出一种强烈的对比反差美，即使你不是球友，也难免跃跃欲试。

187

Haihua International Star Club
海华国际之星会所

The designing conception:
The idea distinguishes from some common assembling public places. The designers combine the idea of resort hotel with artistic museum, then bring in the whole community club, which show a subtle, tranquil and gentle image. The whole picture gives people an intangible beauty scenery.

The lobby:
We choose the cream-color stone to be the fundamental theme, unique totem the emblem, then we adopt lighting decoration and carpet totem pattern to be echoing in the enlarging space, so that an imposing status can be unique presented. The stone for all the wall design are chosen with specially archaized texture material in order to create a gentle atmosphere of the whole space. Furthermore the paintings and sculptures are elaborately matched, with the furnishing layout being of the oriental style, the whole prospect is a

设 计 师：谢启明
设计公司：大企国际空间设计有限公司
项目地点：中国台湾
建筑面积：1617平方米

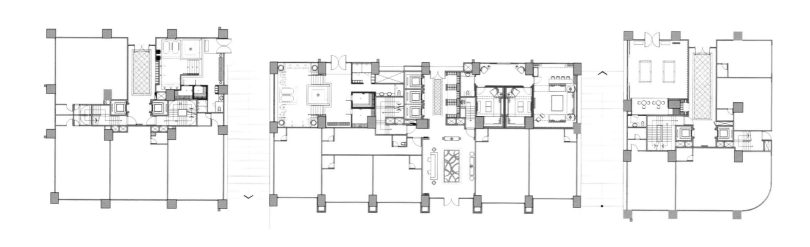

manifestation of elegance, prosperity and humanity.

The café:

The space design is simple and brief, and we choose raw wood and dark glass to be the key theme which creates a crystal tranquil atmosphere. The adoption of large full-height-floor glass can attract the outside green scenery. The coffee machine is self-serviced, which shows humanitarian. Taking a cup of coffee by the window, people can really appreciate the mood of one day and enjoy the static moment peacefully.

The gym:

We utilize simple wooden grid to create a assignment of irregular shape but with rhythm. Also we specially choose mute wooden floor combined with space enlarging sense and fun brought by dark mirror effect, so as to avoid over loud noise. The thoughtful video design make people be fully engross in the sport activities, almost unaware of time elapsing.

The parents-children game room

We ingeniously combined the dynamic climbing area, game zone with the static reading room for parents and their children. So the parents can casually read for themselves while taking care of their kids. Furthermore, we made an exclusive washing room for children, which can reflect our carefulness and uniqueness.

The billiard room

The wooden grid with free figure yields an interesting rhythm in space. The seeming simple design actually comes out through prudent calculation and experiments, such as size of grid, distance and radian measure. The lighting program is connecting with the dark mirror, creating a special atmosphere, which also weaves the beauty of rhythm.

The meditation room

We adopt the wood to be fundamental key, even a pure design conception with gentle illumination design can generates sequence space beauty. The oriental vocabulary is supposed to be the room theme. The suitable dressing room, thoughtful design for putting on shoes and clothes, dressing up, doing makeup, etc. all of these are nothing less than the facilities in top private club.

The spa area

We choose burning surface stones and washing stones for the main material. Given to the safeness and comfortableness, we designed out special lighting effect to form the space conversation among illumination, shadow and water. As to the facility, the area has the steaming room, oven apart from basic facility like water treating pool. In addition, the interior also considerately plans resting room, dressing room, showering room and bathroom for all the visitors.

The KTV room

This part holds three medium and large private boxes and the between porch is the waiting room. The design style is still carrying on brief oriental harmony, but the interior boxes are full of modern LOUNGE BAR feeling. It could be the place where friends have a reunion or the mini cinema. We choose the totem with hollow embossment to go with fashionable furniture inside the boxes, and the dark color stone floor imported from Italy subtly presents a real luxurious taste.

The individual elevator room

In order to keep the style of lobby and the same conception, we still choose the stone material with antique surface, then going with super large 500mm*1000mm quartz brick imported from Italy, which form the main key of the wall. Regarding the ceiling, we use home totem with embossment to match indirect illumination, so that the subtle and elegant space aesthetics can be built. We take good advantage of high-tech connector to install the electric bulletin board of community with cable TV into every elevator room, so that residents will enjoy some superior TV programs like DISCOVERY while waiting for the lift. Meanwhile all different kinds of community information will be caught up on in the shortest time. The mailbox is embedded in the room, which sends out a message of sharpness with enough safety. Even the most specific detail in thoughtful mind is clearly remarkable.

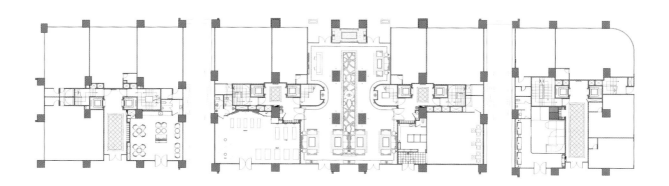

设计概念

有别于一般集合式住宅的公设型态，设计师将本案以渡假饭店会馆结合美术馆的概念，导入整个小区俱乐部，此内敛、沉静、柔和的优雅调性，营造空灵的美感。

迎宾门厅：

以米色系石材为基调，独特的图腾如同家徽，在挑高的空间配合灯光的设计与地坪的图腾拼花相呼应，展现特殊的气势，所有壁面的石材亦特别选用仿古面特殊处理的质感，营造整体空间的柔和感，精心搭配的油画与雕塑等艺术品，及带有东方风格的家具摆设，呈现出优雅大气兼具人文品味的意境。

咖啡厅：

简约的空间设计，以原木及墨镜作为整体空间的基调，产生单纯宁静的氛围，运用大面落地玻璃将中庭的绿意引入眼帘，人性化的自助式咖啡吧，让人悠闲地沉淀心灵，享受静谧的时刻。

健身房：

运用单纯的木格栅，以不规则又带有律动的安排及计划，结合墨镜产生空间感的延伸与趣味，特别选用静音的木地板，避免产生过大的噪声，贴心的影音规划，可以尽情运动流汗，几乎忘了时间的流逝。

亲子游戏室：

将动态的攀岩区、游戏区与静态的亲子阅览室巧妙地安排在一起，让父母可以悠闲地一边阅读，一边照顾子女，里面更是规划了儿童专属的厕所，都是设计师细心的考虑与独到之处。

撞球室：

自由曲线的木格栅，产生空间的趣味韵律，看似简单的设计，却是经过严谨的计算及试验，包括格栅的尺寸、间距、弧度，结合灯光的规划与墨镜的搭配，所创造的特殊氛围，编织出律动的美感。

韵律禅修室：

大量运用实木为基调，单纯的设计语汇，柔和的照明设计，却产生一种秩序的美学空间，东方语汇是本空间的主题，尺度舒适的更衣区，细心地规划出换鞋、更衣、梳妆、化妆等完整的流程，不逊于高级私人会所的设施。

男、女SPA区：

以烧面石材及洗石子作为空间的主要素材，考虑到安全性及舒适感，特殊的灯光设计，营造出光、影、水的对话空间，关于设施方面，除了基本的水疗池各项设施，另有蒸气室、烤箱等设施，也精心地安排休憩区、更衣区、淋浴间及洗手间等完整又全面的规划。

KTV室：

本区规划有三间中大型包厢，内玄关为待传区，设计风格仍延续着简约的东方个性，进入包厢却有着充满时尚感的LOUNGE BAR气氛，可以是亲友聚会欢唱的场所，也可以是小型的电影院。包厢的设计，以镂空浮雕的图腾搭配充满时尚感的家具摆设，意大利进口的暗色系石材地坪，呈现低调奢华的时尚品味。

各栋梯厅：

延续门厅之风格及语汇，同样以仿古面石材搭配意大利进口超大规格500毫米×1000毫米凿面石英砖为墙面之基调，天花板以浮雕之家徽图腾搭配间接照明，营造内敛、优雅的空间美学，运用高科技的接口，将小区电子布告栏结合有线电视安排在各栋梯厅，让住户在等电梯的同时，不仅可以欣赏如DISCOVERY等知性频道的优质节目，同时小区的各种讯息也同时以最短的时间映入眼帘，信箱的设计皆采用嵌入式，大气利落也兼顾安全性，即使最细节的部分也处处包含着贴心的考虑。

Nanning Ronghe Happy House
南宁荣和山水绿城会所

In this city, you go back and forth hurriedly, you have to face all kinds of the bustling, maybe you are tired of bustling in the city what have fixture and concrete only. So you must orderly to tranquil's desire degree very imperative.You don't need to find deliberately, because there are such a group of people in mind to help you guard patch of quiet forest, let you listen to gurgling water and chirping birds.

Guangxi Ronghe Happy House in north of Nanning Mingxiu Road, look around and stacked green belt, wide landscape view. About clubhouse, choose modern luxury style show her location: dignity and fashion, also they both have fine modern building techniques and materials with art display, showing a natural, elegant high-end clubs.

In this design, the designer trends to lead the exterior environment and the architectural style to the interior design idea, which is close to nature, sun and water. The diagonal element of the architecture which

设 计 师：戴勇
设计公司：戴勇室内设计师事务所
项目地点：中国广西
建筑面积：3600平方米

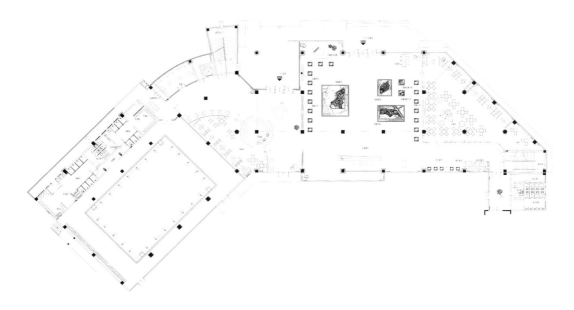

is simply shows the elegant, beautiful, peaceful and spiritual artistic conception. The three things are shown in the design that are the space arrangement, the light design which is full of feeling and the thematic art decorate. This design makes the space convey a warm and noble temperament, which lets people enjoy it so much as to forget to leave.

在匆匆往返的都市里，面对熙熙攘攘的种种，你厌倦了在钢筋混凝土城市里的奔波忙碌，你对静谧的向往程度如此迫不及待。你不必刻意去寻求，有这样一群人在帮助你守护心中的那片宁静森林，让你听潺潺流水与啾啾鸟语。

广西荣和山水绿城地处南宁明秀东路北侧，环顾四周，绿化带叠起，景观视野广阔。作为项目住客会所，选择现代奢华风格来突显其尊贵、时尚的定位，既具备现代建筑手法又兼有细腻的物料搭配及艺术品陈设，呈现出一个自然、优雅的高端会所。

在设计时，设计师有意将外部环境及建筑风格引入室内设计理念，亲自然、亲阳光、亲水景，建筑的斜线元素的延续，手法简练大气，表现出优雅、唯美的姿态，平和而富有内涵的气韵。建筑空间的处理手法、富有气氛的灯光灯饰设计及主题性的艺术陈设，一直贯穿整个空间设计的始终，让空间传达出温馨尊贵的气质，让人流连忘返。

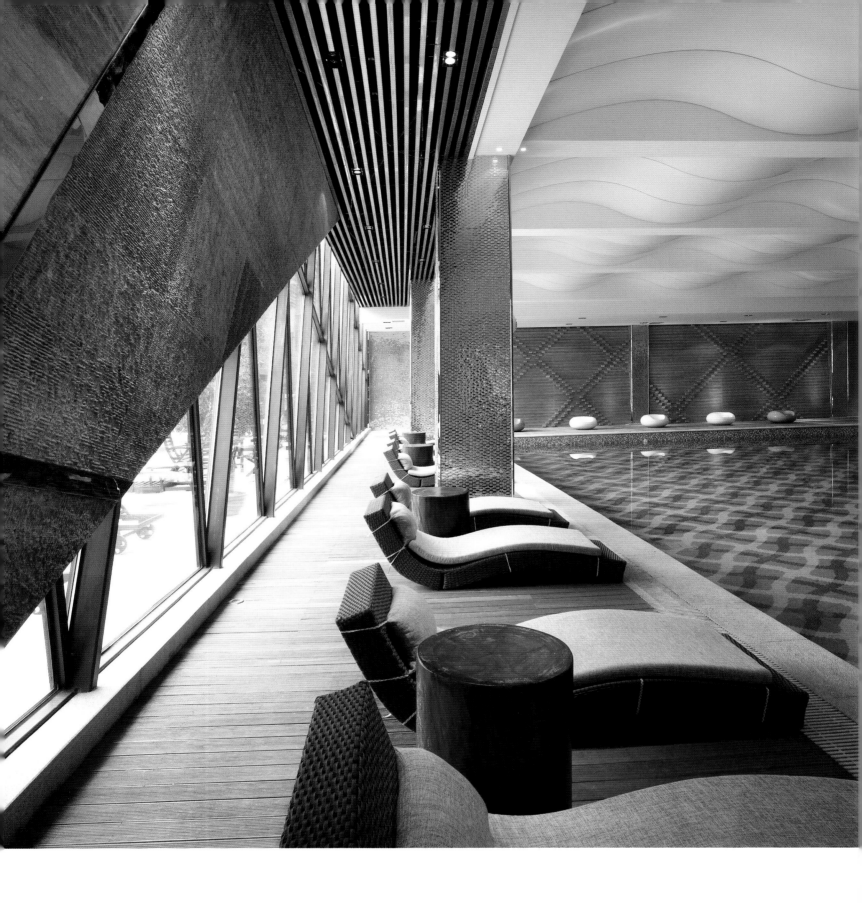

Ronggui Royal Sea Club

美的容桂御海东郡会所

"Royal Sea Club", what a poetic and elegant name! As a famous club design office in Foshan city, we start this case around the leisure idea of southeastern Asia sea weather, meanwhile adding to luxurious and classic sense. Now, let people get rid of one day's fatigue and fully enjoy this leisure scenery.

Honorable but not showing off, the lobby design can tell this suitable and charming life style, integrated and upright. The ceiling design is echoing the space subtle separation, which bring about the vision focus. Most of the space adopts natural daylights, so the interior sight can penetrate with exterior sight, which will play up an elegant environment. However, some individual spaces like the bar, the cigar room and reception room, etc., embrace a modern gracious style. The stainless steel wall painting with vivid color tone, the texture wall paper, the floor in old way dealing, the crystal decoration and the perfect furniture layout etc., all of which endow the space with infinite vitality.

设 计 师：区伟勤
设计公司：广州市韦格斯杨设计有限公司
项目地点：中国广东
建筑面积：1785.15平方米
主要建材：大理石、木地板、墙纸、瓷砖、地毯、布艺、马赛克、黑镜、清镜、镜钢、砂钢、灯片

御海东郡，一个富有诗意而又典雅的名字。作为佛山市知名楼盘的会所，整体围绕东南亚滨海气候的休闲写意展开设计，同时富有豪华气派与经典，现在，就让大家从一天的工作劳累中解脱出来，去感觉这种休闲写意吧！

尊贵光鲜而不张扬，从大厅设计可以看到这种恰如其分的舒适典雅生活典范，方方正正，天花设计呼应空间隐性划分并带出视觉重点。大部分空间则运用自然光线、视线与室内外相互穿透，渲染出优雅的空间气氛。而在酒吧、雪茄吧、会客厅等个性空间，则有典雅时尚感，不锈钢、鲜明色调的墙身、纹理墙纸，木地板做旧处理、水晶饰品点缀、家具完美的陈设布局等均赋于空间无穷的活力。

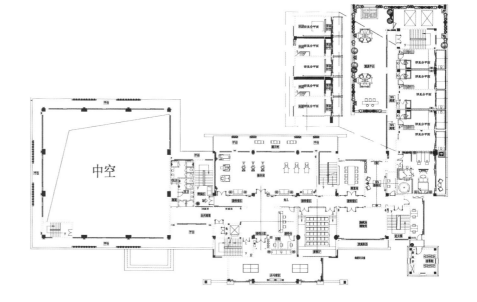

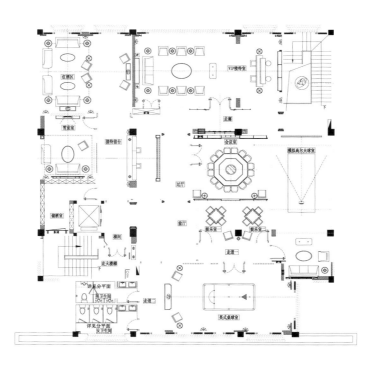

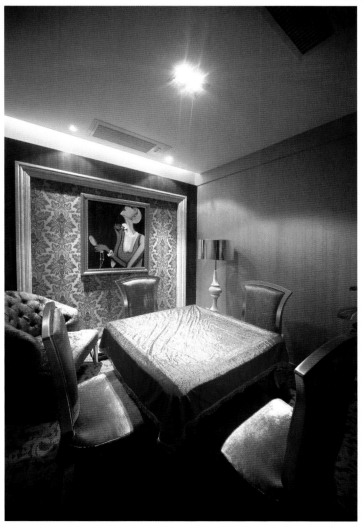

Shenzhen Vanke Tangyue Club

深圳万科棠樾会所

This case is of Chinese style, pursuing the beauty of symmetry and balance of space, which can get the Chinese charm and taste.

In the arrangement of space, designer simplifies the decorative elements to the most in the background of Chinese themes, and creates a space where the air can flow freely, which can avoid the sense of oppression and can meet the functional feelings which this club intends to express.

In order live with the life concept of eco-compatible livable space of the building, the space is mainly decorated with the solid wooden furniture, and dotted by green plants. Besides, there're functional areas like chess room, KTV in this club, which can meet the needs of the residents.

设 计 师：韩松
设计公司：深圳市昊泽空间设计有限公司
项目地点：广东东莞
建筑面积：3000平方米

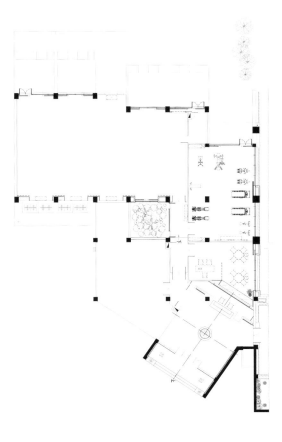

本案设计定位为中式风格，努力追求中式的对称美和空间平衡关系，达到中式的神韵和气质。

在空间布局上，在中式主题的背景下，设计师尽量简化装饰元素，力求打造出一个大气流畅的空间，避免使人产生压抑感，以符合会所所要表达出的功能感受。

为了与楼盘宜居生态生活理念相符合，空间以实木家具为主，并装点绿色植物。另外，该会所还设置了棋牌室、KTV等功能区，以满足居住者的需求。

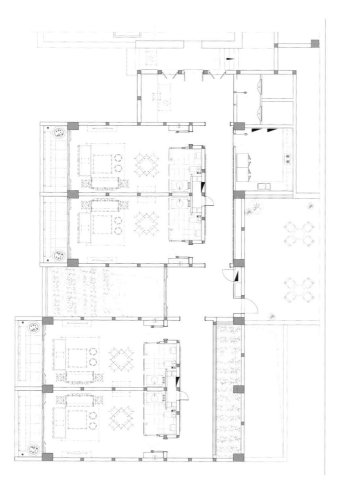

Felicity Famous Kunlun Business Club
富临名家昆仑商务会所

Kunlun Business Club is a comprehensive hotel space which provides high-end catering, entertainment, guestroom service and meets the needs of high-end consumption. The overall design ideas continue and enhance the core cultural philosophy of Maxforum group: blessed by riches and honor, style bearing artistic distinction. Peony, flower of riches and honor, is used as the basic image for design elements, which is integrated into every space and detail through rich, subtle design vocabulary. Among which, the most classical and notable symbol of peony is the huge piece of peony art work found at the reception background of the first-floor lobby and lounge background, becoming the distinguishing focus for the whole club. Mixing luxurious elements from both classical European style and contemporary Chinese

设 计 师：史新华、黄建、王志刚
设计公司：杭州易境环境艺术设计事务所
项目地点：中国浙江
建筑面积：10,000平方米
主要建材：金蜘蛛大理石、黄金洞石、花樟木钢琴漆面、手绘墙纸

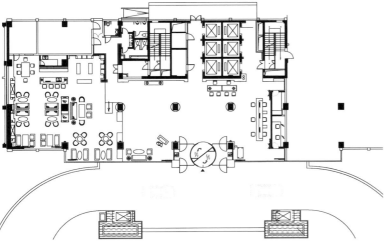

cultures, by the means of space, furniture, lamps, works of art, etc, and through special artistic mix and match, the overall design style produces a strong stylish, classically sumptuous characteristic of culture, and especially the design of lightings resulted from an implicit cooperation with professional consultant gives the space the depth of visual aesthetic provoking endless thoughts.

富临名家昆仑商务会所是以提供高端餐饮、娱乐、客房服务功能，迎合高端消费需求的综合酒店空间。整体设计理念延伸并提升了富临名家集团核心文化理念：富贵临门、名家风范。设计元素以花开富贵的"牡丹"为基本意象，以丰富含蓄的设计语汇融入各个空间及细节之中，尤其以首层大堂的接待背景、大堂吧背景的巨幅牡丹艺术品最为经典，成为整个会所标志性亮点。整体设计风格融合了欧州古典和现代中式文化中的奢华元素，以空间、家具、灯饰、艺术品等方式，经过特别的艺术混搭，形成一种具有强烈时尚和经典奢华的文化性格，尤其在灯光设计上经过与专业的顾问默契配合，使空间具有了耐人寻味的视觉美感"厚度"。

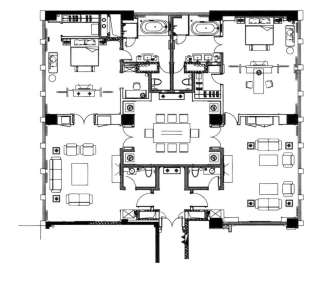

Sam-fong Renfa Constructions Reception Center

仁发建设三峰接待中心

The reception area emulates an art gallery to bring art into life and create an environment filled with culture.

This case is the Sam-Fong Reception Centre for Renfa Constructions in Hsingchu, situated on Ziqiang Road, Zhubei City. Two of the centre's sides are facing the streets, where slanted sides, cuts, glass and concrete structures have been used to create a building that resembles a work of art. The red entrance stems from the concept of a framed view. The glass structure is set inside the architecture itself, allowing viewers to catch glimpses of the interior. This method was also used in the design of the vertical planes. The walls of the discussion and office areas have linear openings

设 计 师：朱柏仰
参与设计：蔡宜芳
设计公司：暄品设计工程顾问有限公司
建筑面积：277.2平方米
主要建材：大理石、喷漆、墨镜、磁漆

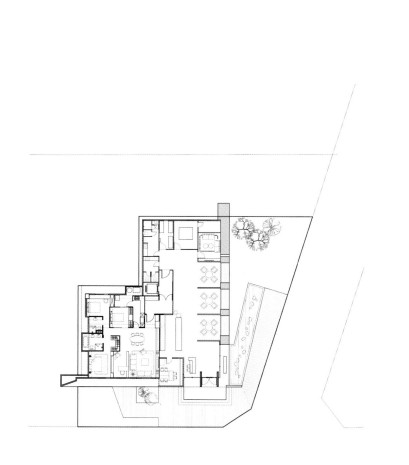

that act as visual guides for visitors, so they can see the waterscape and greenery outside interacting with traffic and pedestrians. Around the building, elements inspired by the lotus flower are used in the landscape design. Sculptures resembling lotus flowers stand elegantly in the pond, whilst the ripples and waves, the light and shade, and the reflections of the sky and the environment create a bond between art and nature.

The interior design also borrows from art galleries and consists mainly of plain white walls - the best backdrop for displaying artwork. We have collaborated with the Taiwanese artist Mingsheng Lee, whose series "Our Mother the Mountain Chains" is exhibited here. The ceiling also uses a flowing design to create rhythm, echoing the soft ripples of the water outside, whilst the multi sided decorative shields with their slight metallic sheen connect the entire space. Gradual changes, twists and light-plays lead the spatial transformation of the reception centre.

以提倡生活带入艺术，创造一个充满人文气息的环境，而将接待中心打造一个充满艺术感的画廊空间。

本案为仁发建设"三峰接待中心"，坐落于竹北自强路上；接待中心两面沿街，以斜面、切口、玻璃结构与混凝土结构的搭配，塑造出如艺术品般之建筑；红色入口以一种"框景"之构想，将玻璃结构嵌入建筑体，进而窥视室内空间，此手法也用于建筑立面之设计，洽谈区与办公室墙面以带状开口引导人们视觉走向，能望出户外水景和绿意以及车潮与人潮之流动。建筑周遭环境塑造以大自然中的"荷花"之元素艺术化，设计如水中荷花之雕塑一朵朵站立在池中，而水波、波光、光影映射天空与环境，营造出自然与艺术相融合的感觉。

室内空间借鉴画廊的设计手法，以素静白色墙面为主，作为呈现艺术画作最好的空间场景，与台湾艺术家李铭盛合作，将其画作［山脉——我们的母亲］系列在此展示；天花设计也以一种流动之韵律布满天花，呼应了户外水波曲线，带有金属光泽的多边形片状饰板贯穿整个空间，渐变、曲折、光影主导了接待中心空间变化。

Shunde Coast Constellation Club
顺德海岸星座会所

This case is defined as sporting club, based on management. With the view of business operation, we choose "round" "white" for designing element in order to attract visitors' attention. The space can create a strong "aura" that subtly controls the customer's interests, so that it can please people's mind. And they will linger in the room as possible as they can, therefore the business goal will be reached.

First of all we set up a "receiver" right at the entrance, including the function of music enjoyment, internet, TV and the resting room, etc. The other side of "receiver" is facing the outside garden scenery, so the role of "receiver" is playing for the waiting room across the lobby. Be in this small space, people will always indulge themselves in pleasure that they forget to leave. That exactly is the effect brought by the aura of "round gold".

Then we replace "golden" by "white" in other spaces, which also can reach the same effect. However, if we all adopted the element of "round" and "white", obviously it would come up with an unbalance situation. After all this is a healthy sporting club, the non-balance does no good to the visitors, so we prefer the fruit green color which symbolizes "healthy, vitality, ecology" to be the other theme color of the room. Since green belongs to "wood", and "white""green"

设 计 师：区伟勤
设计公司：广州市韦格斯杨设计有限公司
项目地点：中国广东
建筑面积：1800平方米
主要建材：白色人造石、黑石、果绿色氟碳漆、清镜

actually perform an effect of "wood will lose in the battle between gold and wood", we need "water soothes the conflict", and black stands for "water". Finally three of them can bond a dependant relationship: wood comes from water, which derives from gold, so that a harmonious situation will be generated. It is a balance as to vision; it is a kind of relaxation from the opinion of customers, and the operator rather think it as a circulating stability.

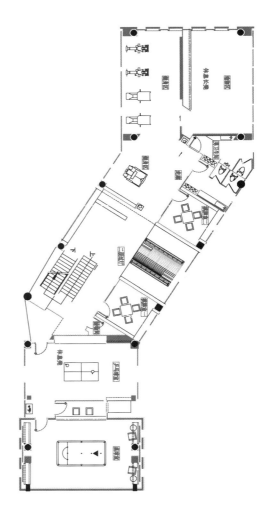

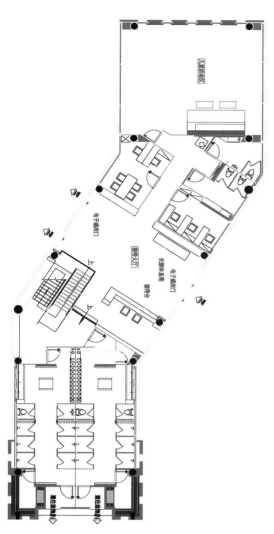

本项目定位为运动型会所,以经营为本,基于商业运作的考虑,吸引顾客为目的,固在设计元素上取用"圆""白"。令空间产生强烈"气场",以气制客,水来相生,从而可以取悦于顾客的眼球,让客人多停留于此空间内,以达到商业目的。

首先我们在入口处设定一个圆形的"听筒",筒内设有音乐欣赏、网络、电视、休息等功能,"听筒"的另一端开口正对着会所以外的园林景观,此"听筒"体为过厅的等候区,在这小小天地里,可让客人有流连忘返的感觉,这正是"圆金"气场所发挥的作用。

然后,如果我们也在其他空间上取"白"代"金",同样也会起到相同的作用,但全为"圆"和"白",显然是一种不平衡状态,也不和于客人,毕竟是一间健康运动型会所,所以另采用具有"健康、活力、生态"意义的果绿色作为空间的另一大主打色,因绿色属"木","白"和"绿"实为"金木相伐,木遭殃",还需要"以水相济"调和,而黑色属"水",最终三者连成相生关系:金生水,水生木,以达到协调状态。而从视觉上则是一种平衡,从客人角度理解则是一种精神的放松,从经营者角度理解则是一种流通的稳定。

Beijing Xiangsu Property Broker
北京像素售楼处

Appearance

The appearace was formed of many separate tubes accumulated. It includes four kinds of colors, they are white, light grey, dark gray and black. Overlapping and accumulating so as to produce bump sense.

Sales center hall

Sales centre hall is a very important part of sales centre used as the whole project sale centre, including negotiation room, guest rooms and so on.

The hall, including enternce hall, seeks to pursue a sense of luxury and dynamic space to Interior decoration also developed the cell concept. The wall accumulate different length and gloss wooden cases, 400m㎡*400 m㎡, to get bump and layer sense. The roof is created by 400 m㎡*400 m㎡ droped tubular lightings. Its borders are made of Japanese paper. Then the ground integrate outdoor and indoor, integrate outside landscape to indoor.

设 计 师：迫庆一郎、最上有世、栗本贤一
设计公司：SAKO建筑设计工社
项目地点：中国北京
建筑面积：3000平方米

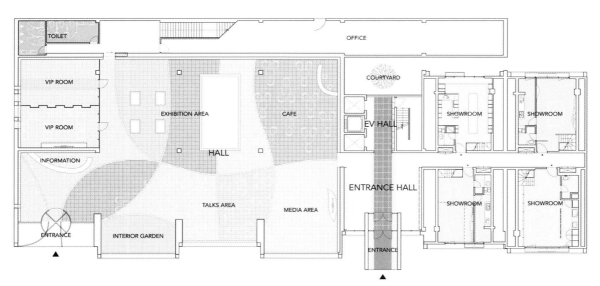

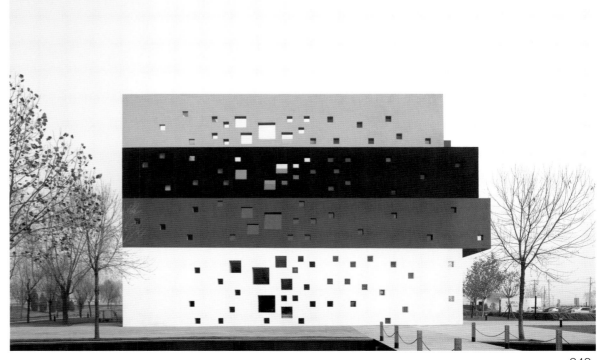

外观

外观是由多个各自独立的管堆积而成。管分为白色、浅灰色、深灰色、黑色四种，各色管长度不一，重叠累积，从而产生凸凹感。

销售中心大厅

销售中心大厅是用做"北京像素项目"整体销售的场所，包括洽谈室、贵宾接待室等，是销售中心非常重要的组成部分。包括来宾入口在内，整个大厅设计都力求追求一种与洽谈室相称的豪华感和震撼力的空间。

内部装修也意识"北京像素项目"整体细胞观念，墙壁部分累积400×400的正方形木箱，各个木箱拥有不同的长度和色泽，创造出凸凹层次感。顶棚部分由400×400的边框状的日本纸构成的灯筒从顶棚垂钓下来，起到照明作用。

地板部分注意内、外的一体性，是将室外园林风格直接引入室内的一种风格。

Chengdu Languang Hejun 47 Mu Sales Center
成都蓝光和骏47亩售楼中心

The '47 Mu' is a real estate project for young and middle-aged white collars. The concept is 'SOFA Music Community'. The designer relevantly chose 'music' as the main theme to express the whole space, considering the understanding of music as the start of design. In his opinion, music is an intangible and abstract thing. Different music has different rhythm and mood. The main point of this case is how to transfer the beauty of time into beauty of space and express the pursuit of music. The design of this project is balmy and bracing, which combines people's longing for nature and the rhythm of music. Limited by cost of construction, this case adopts the most common way of pieces to create a young and robust 'Green Music Community'.

设 计 师：殷艳明
参与设计：沈磊
设计公司：深圳市创域艺术设计有限公司
项目地点：中国四川
建筑面积：约500平方米
主要建材：橡胶地板、肌理涂料、软膜天花、人造石、仿云石灯片

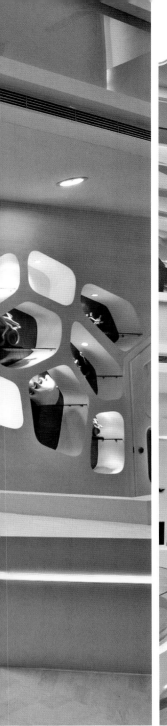

"47亩"是面向中青年白领群体的地产项目，经营概念定位为"SOFA音乐社区"。作为该楼盘的售楼部，设计师相应地把"音乐"作为空间表达的主要概念，把对音乐的理解作为整个设计理念开始的起点。在设计师看来，音乐是一种无形、抽象的事物，不同的音乐具有不一样的节奏感、不一样的情绪、不一样的韵律。如何把音乐这种时间美的特点转化为空间的美，让整个空间具有完整、起伏的旋律，表达一种对音乐主题的向往和追求，是本案的主要着力点。本案的整体设计张弛有度，把人对自然的美好愿望和音乐旋律巧妙结合在一起，由于造价所限，本案用最普通的材质通过体块穿插的手法营造出一个年轻而又充满生命力的"绿色音乐社区"。

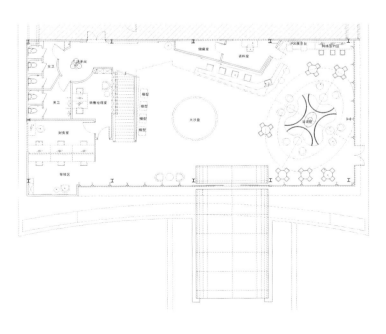

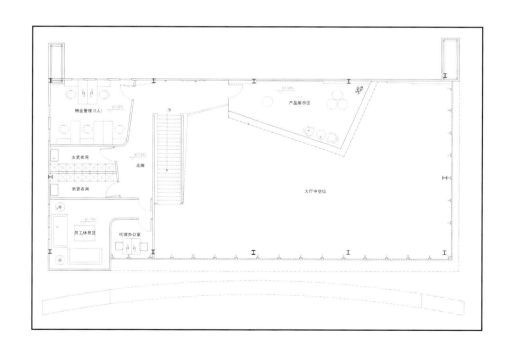

Oriental Jinshi Reception Center

东方金石接待中心

Oriental Jinshi reception center is located at Zhuhai east road in Jiaonan, Qingdao. Customers are mainly mining bosses, whose social value and status are different from ordinary people. The base position also has good natural seaview, so the case has quite an index in this area.

Customers in this case have the background of mining, hoping to create the best reception centre in Jiaonan gulf, outdoor algorithm at the first of case changes into indoors. After analyzing and discussing the base, dividing into several concept area, reception, receiving visitors negotiating area, signing and office working area, the base in this case abuts the gulf, is located in 29F and has a good vision and ventilation.

Reception center is mainly commercial business behavior, contains activities such as meeting, explanation, banquet, etc. Moving lines must consider people passing by and blends into the outdoor natural landscape.

The algorithm has a large core tube and wants to use the core tube to form a modelling of stereo feeling. Study many materials, such as granite, metal plate, composite materials and try to coat with them, which cannot appear luxury ore feeling. So it is adopted gold foil fabrics, firstly, cover special coating on metope, press out texture and finally

设 计 师：黄士华、孟羿彣
参与设计：藏弄设计团队
设计公司：隐巷设计顾问有限公司
项目地点：中国山东
建筑面积：1400平方米
主要建材：金箔纸、银箔纸、镜面不锈钢、胡桃木、夹胶玻璃、金属马赛克、小牛皮皮革

cover with goldfoilpaper, directly and clearly show the concept of "bronze".

The whole design materials are mainly consist of modern materials, by continuous circle element, shaping Oriental feeling. Around the core tube, use 4 mm solid wiredrawing stainless steel laps to frame it. Use the contrast of rough and meticulous, light and dark, tradition and modern, producing macroscopic dimensional feeling, use mixing material forms to strengthen fusion of Jinshi and Oriental concept.

Major moving lines use crystal glass chandelier as segregation. Light goes through crystal glass, brings dazzle light and increase costly feeling. The original elevator hall on both sides of the aisle, makes a mobile terminal scene wall, in addition to unified customer visiting moving line, it can avoid customers leaving the reception center because move line is too simple, while it has safe effect and can be locked at night for protection.

Model's location is at the entrance of the gate, the background can overlook the seaview, without a break, make visitors know product superiority at the first time. Left wall makes the space produce administrative level feeling, and also serves information. The right side is reception area, different white marble cutting combination. It's also an extension of bronze concept.

This case uses many uncovered compartments made by copper strips. Put plants' modeling, beautiful and privacy. In the reception service area, setting wall concept is the same as the core tube and combined with Chinese bamboo modeling.

Negotiating and receiving visitors' area raise the ground, making customers can appreciate seaview while sitting on the sofa, use different ground materials to define space function, and configurates a bar area, which can be used as the function of banquet and catering. In the future, the reception center can transform into a club. The second using function is listed in the plan at the stage of scheme design.

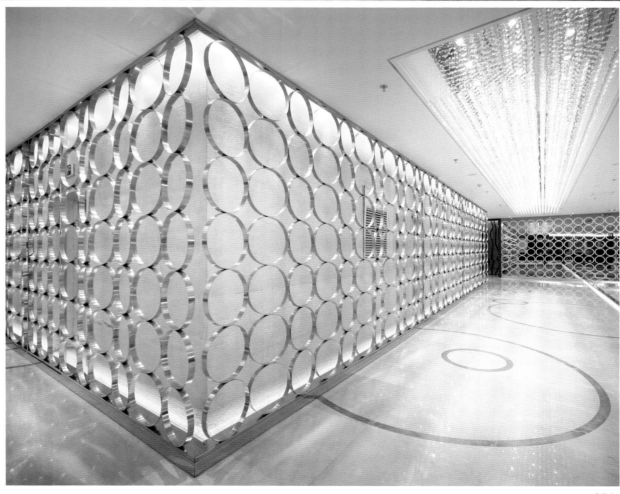

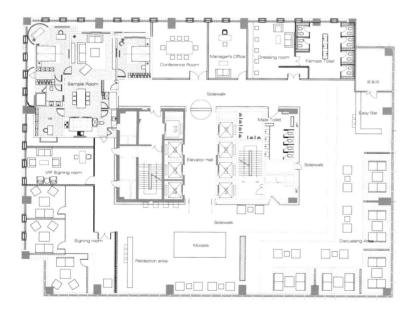

东方金石接待中心位于胶南市珠海东路，主要客户为矿业老板，其身价与地位较高，基地位置同时拥有良好天然的海景，故此案于该地区具有相当之指标性。

本案开发商具有矿业的背景，希望能打造出胶南海湾最好的接待中心，方案最初由室外建筑体转变至室内，基地经过分析与讨论后，划分出几个概念区域。接待区、会客洽谈、签约与办公区，此案基地紧靠海湾，位于29F，视野与通风良好。

接待中心主要是商业买卖的洽谈场所，还举办会议、说明、宴会等活动，整个空间也必须考虑人流环绕，并融入户外之天然景观。

建筑体有一大型核心筒，想借由核心筒形成具有体量感之造型，研究许多材料，如花岗岩、金属板、复合材料等尝试包覆，都无法呈现奢华矿石之感，故采用金箔面料，先于墙面上覆上特殊涂料，压出纹理，最后再覆上金箔纸，直接明确地表现出"金石"概念。

整体设计材料主要以现代感之材料构成，借由连续性圆圈元素，塑造东方感，核心筒四周以4毫米实心拉丝不锈钢圈框住，利用粗犷与细致、明与暗、传统与现代之对比，产生巨观之空间感，借由混合材料的形式强化金石与东方概念的融合。

主要动线以水晶玻璃吊灯作为区隔，灯光透过水晶玻璃产生之炫光增加奢华感，原有两侧信道之电梯厅，制造一座活动式端景墙，除了统一客户参观动线，可避免客户因动线过于简便而离开接待中心，同时具有安全效果，夜间可上锁形成防护。

模型台位于大门入口，背景为可眺望之海景，一气呵成，使参访者能于第一时间了解产品优势，左侧之墙体为使空间产生层次感而存在，同时也是信息墙，右侧为服务接待区，不同白色理石切割组合，同样为金石概念之延伸。

此案采用许多非遮蔽性之隔间，由铜管制成之隔间，放入植物的造型，美观与隐私兼顾，接待服务区背景墙概念与核心筒相同，并融合中式竹林造型。

洽谈会客区将地面抬高，为了让客户坐在沙发上同样能欣赏海景，透过不同的地面材料界定空间功能，并配置一吧台区，能同时兼有宴会、餐饮的功能，未来接待中心可转型成会所，二次使用的功能于方案阶段设计时即已列入规划。

Green Generation

陆江当代

Through cant exterior wall models different angles' perspective result, it uses system earthquake steel structure as the standpoint of architectural quantity body, also as a part of the outdoors device sculpture, decks building's drama tension.

The first floor is the entrance and parking lot, through the first floor reception, go upstairs to the second floor's reception hall, it uses cant turning, makes incision exterior wall fold to interior wall, forms a twin architectural element. Indoors part uses board to increase its narrow window detail and makes the person observe 3D effect of Yin, Yang face from different angles, when western sunlight goes through the narrow window and reflects on the ground and metope, it will constitute a sequence of uncertain beam as time changing.

Like a L pattern division is divided into negotiation area, model district and sample room. The display stage of model district continues custom-made furniture stage on vestibular, which will make scale amplification as a architect's work table or a building device exhibition in an art museum.

设 计 师：甘泰来
设计公司：齐物设计事业有限公司
项目地点：中国台湾
建筑面积：1745平方米
主要建材：氧化镁板、风化木、黑铁、强化玻璃、秋香木木皮、茶色玻璃、茶镜、黑色洗子、烤漆、皮革

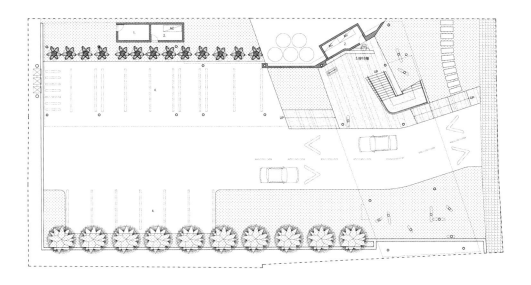

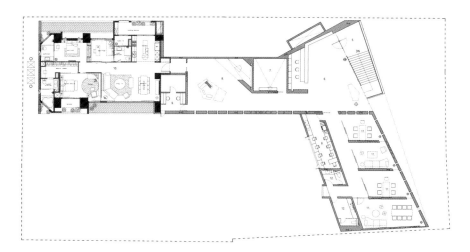

　　透过斜面外观墙塑造出不同角度的透视效果，并利用制震的钢结构作为建筑量体的立足点，同时也成为户外装置雕塑的一部分，装点出建筑的戏剧张力。

　　一楼为车道的出入口和停车场，经由一楼接待柜台从楼梯而上，进入二楼的接待门厅，运用斜面形式的转折，使切口外观墙翻折到内墙，形成一体两面的建筑元素，在室内部分并以木板来增加其窄窗细部，使人可于不同角度位置观察到阴、阳面的三维效果，每当西晒的阳光透过窄窗倒影在地坪、墙面，都会随着时间的变化而构成序列不定的光束。

　　有如L形的格局划分，则分为洽谈区和模型区、样品屋。模型区的展示台面延续门厅所见的订制家具，将尺度放大，有如建筑师的工作桌或是美术馆内的一个建筑装置展品。

HYL Gallary Independent Architecture

HYL Gallery 自主建筑创志馆

HYL gallery is a series of different proportion of V-Shape type space. From appearance to inside, the ground to the top level, compressing or done, turning or climbing, organized into shape. A provided different angles and height's Viewing Device, at the same time, more breakthroughs the form of traditional sales department. Use the private art museum's appearance. Perform the space stories of three architects, conveniently link to the base's infinite and priceless excellent landscape.

设 计 师：甘泰来
设计公司：齐物设计事业有限公司
项目地点：中国台湾
建筑面积：1F 570平方米、2F 32 平方米
主要建材：石材、钻泥板、木皮、灰镜、epoxy

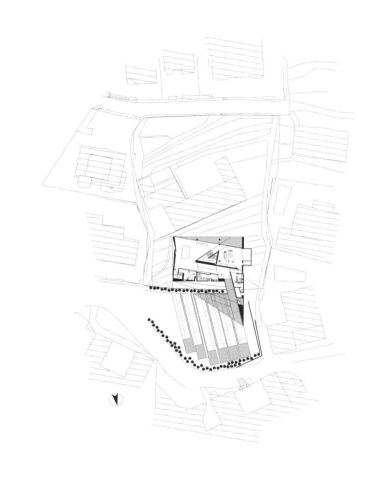

HYL gallery由一系列不同比例之V形空间，自外形至内里，再从地面至高层，或压缩或伸张，或转折或攀升，形塑组织而成。在作为一可提供不同角度及高度之观景装置（Viewing Device）的同时，进一步突破传统售楼处的形式，以私人美术馆之姿，将三位建筑师的空间故事展演，顺势链接至基地无限与无价的极致景观。

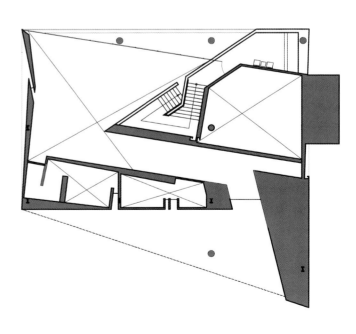

Forest Court

森林苑

The whole building adopts L configuration, echoes base's diagonal forest park, reuses weathering fold wood board level modelling. It metonymies natural forest imagery, collocates white stereo L architecture at the edge of building. Shape highlights L agile city sense of half outdoor entry hallway, then opens the next series of space trip, besides continues weathering wooden modelling fold board wall indoors by using a series of different radian and curvature gradient parallel plate, it brings "white clouds" or "ridge" image. Weathered wooden modelling wall is sunlight filters, according to different conditions of internal space, collocates stitching openings, lets sunshine penetrate into indoor, which makes sunlight penetrate into internal space, forms the dynamic lighting with different time like time mark and notes on the "cloud" and "ridge".

设 计 师：甘泰来
设计公司：齐物设计事业有限公司
项目地点：中国台湾
建筑面积：1F 330平方米、2F 100平方米
主要建材：风化木、茶色玻璃、茶色镜面、墨镜、黑色洗石子、白色喷漆、抛光砖、皮革、银白龙石材

整体建物采用L形方式配置，以呼应基地斜对角的森林公园，再利用风化木折板层次造型，转喻自然森林意象，搭配着白色立体"L"形架构于建物边缘，形塑凸显出半户外入口门厅，进而开启接下来一系列的空间之旅。室内除了延续一体二面的风化木造型折板墙外，亦利用一系列不同弧度及曲率渐变的白色平行板，带入"行云"或"山脊"之意象。风化木造型墙是阳光的过滤器，依据空间内部不同状况，搭配不同疏密的开口，让阳光透入室内后，会随着不同时间形成动态光影，恰似时间的印记，注记于"行云"与"山脊"上。

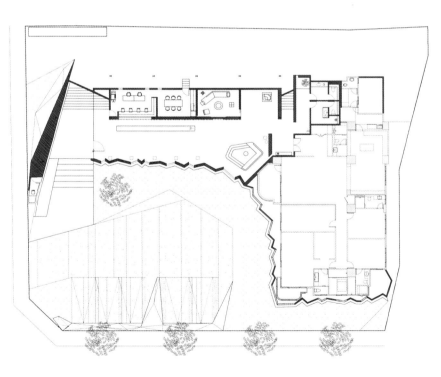

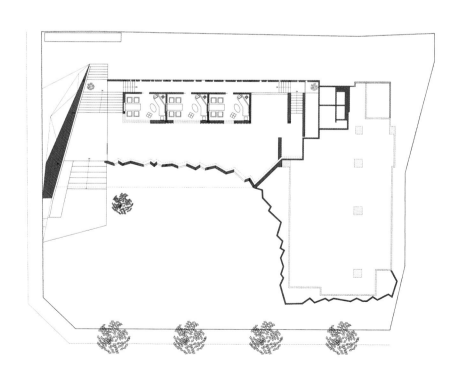

Greentown Real Estate Zhuji Sales Department
绿城房产诸暨售楼部

The reforming case of Green City of Zhuzan old town is a large urban complex that has been working with Zhuzan government. The total area reaches 170 thousand square meters, including super shopping center, five-star hotel, commercial pedestrian street, business office, high-class apartment with scenery and mountain villa in the downtown, etc. As the case proceeds, the old temporary sale department has already not satisfied the need, so it takes granted that the new permanent sale building will emerge.

The whole sale center covers 1700 square meters sale department, 1300 square meters office building and 2 suits molding rooms. In order to follow the comprehensive defining demand given by Green City group, the design style

设 计 师：陈海文、吴悦
设计公司：杭州郑陈设计事务所
项目地点：中国浙江
建筑面积：4000平方米
主要建材：金世纪、凡尔赛金大理石、欧式暗花壁纸、金银箔纸、镂花灰镜、黑镜、石材马赛克、定制黑檀木皮

is orient to modern European style. Given that different demand in different region, we set up some varieties on the style expression. For instances, the sale department speaks out luxury and grandness, presenting an imposing feeling of the most superior estate; the office building comes out with a brief and simple presentation, with a prudent and subtle style; the model room also is elegant and warm, which not only coincides with the estate orientation but also fulfills the owner's desire for home.

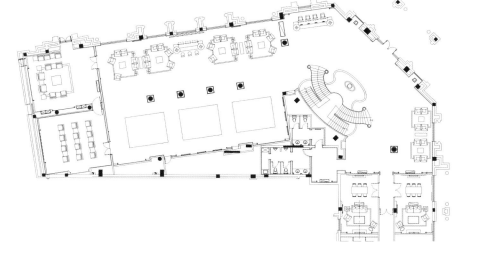

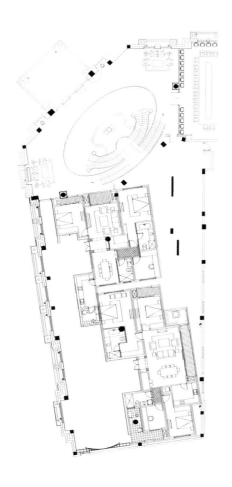

绿城诸暨旧城改造项目是和诸暨市政府合作的一个大型城市综合体，总建筑面积达170平方米，包含有大型购物中心、五星级酒店、商业步行街、商务办公楼、高级景观公寓、市中心山体别墅等物业类型。随着项目的进展，原来的临时售楼部已经变得无法满足需要，新的永久性售楼中心也就应运而出。

该售楼中心总建筑面积包括1700平方米的售楼部、1300平方米的办公楼和2套展示样板间。配合绿城集团对项目的综合定位要求，整体设计风格为时尚欧式风格。考虑到各部分功能分区的不同要求，风格表达上做出一定的差异性：售楼部更为豪华大气，体现诸暨最高档楼盘的气势；办公楼整体表达简洁明了，风格特征点到为止；样板间典雅温馨，既吻合楼盘定位，同时又能满足业主对家的渴望。

Guangzhou Chuangyi Yayuan Sales Center
广州创逸雅苑售楼中心

Designed to the standard of five-star hotel lobby, the whole sales space presents us a classical style of metropolis. In the process of the design, designers not only pay attention to fine details, but also display this sales space in a unique way. Whether external space or internal space, all reflect the fun of creativity. The reception area, exhibition area, negotiating areas, audio-visual area, and the subscription and contract area make full use of the space, from beginning to end without stopping.

The exhibition area is located in the most prominent position in the center. And the working place for the sales staff is in the hall. The shape of the information desk is simple yet stable, with a combination of red and white. The sign of

设 计 师：王哲敏
设计公司：诚之行建筑装饰设计咨询有限公司
项目地点：中国广东
建筑面积：1450平方米

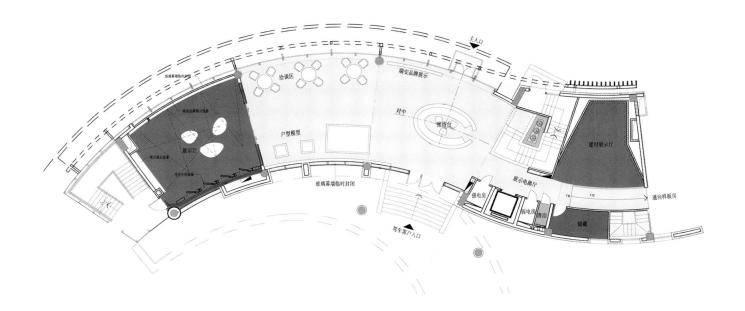

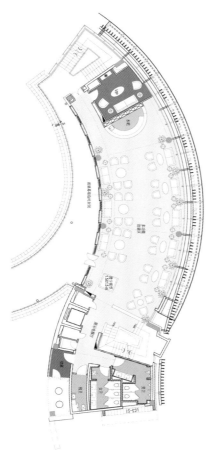

Rui'an creation is inlaid in the front of the desk, highlighting the corporate image and brand of the sales department.

Behind the desk is the audio-visual area. The carpet with yellow rings, white PPT curtain, and the lighting decoration with a look of the stars, all of these will move the visitors.

In the negotiation area, the design of the ceiling do not adopt complicated forms, but use a circular ceiling to emphasize the theme, which makes the space more colorful. Each sofa district, which is relatively independent in one region, gives the guests a feeling of having a more distinguished position.

The sales department, as an important part of the whole real estate, is not only completely different from the current monotonous sales departments, but also reflects the taste of the real estate. The designers take into consideration of make full use of this place to make a sales department and also a leisure place, which displays a leisure yet luxury space, highlighting the modern and magnificent effect.

整体空间呈现出都会经典风，以五星级酒店大堂的设计标准来打造售楼处。设计过程中，设计师不但注重精美的细部，而且以独特的方式展示了这一销售空间，无论是外部空间还是内部空间，都体现了妙趣横生的创意。接待区、展示区、洽谈区、影音区、认购签约区一气呵成，合理利用了空间。

展示区设在中央最显眼的位置，销售人员工作区设在大厅，服务台造型简约，红与白相结合，简单又不失稳重，"瑞安·创逸"镶嵌其中，突出了售楼处的企业形象与品牌。

在服务台背后是影音区，黄色圆环的地毯、洁白的PPT幕布、犹如星光的灯饰定会使看楼者心动！

在洽谈区，天花板处理上没有烦琐的天花造型，而是采用圆形吊顶来强调中心主题，使空间更加丰富多彩。在每个沙发区，相对独立的一块区域，客人处在较尊贵的位置。

售楼处做为整个楼盘的重要组成部分，不但要完全不同于当前过于单调的售楼场所，更要体现房地产商的品味，设计师把它定位于一个售楼部和休闲场所两用来考虑，体现了休闲与豪华的都会空间，突出了现代、宏伟的效果。

Wulinfu Sales Center

杭州大家武林府售楼中心

The design of this case abandons the conventional design style. While the imagination of space is fully showed in the layout of the club, meanwhile, the theme of science and technology is also integrated into the design, making the whole club be full of modern flavor. The use of diamonds and totem patterns enriches the sense of hierarchicy and cubic effect of the space. Delicate and elegant flexible decorations with crystal lamps carefully create a luxurious and elegant commercial space so that every customer can get a honourable and comfortable feeling.

设 计 师：徐少娴
设计公司：Gotomaikan International Limited
项目地点：中国浙江
建筑面积：约328平方米

本案设计摒弃了常规手法，以流畅的空间想象力贯穿整个售楼中心，同时融入科技感主题，使它无疑浑身上下都充满现代气息。大量运用菱形与图腾花纹，丰富空间层次感与立体感，精致典雅的软装加上水晶灯的搭配，精心营造出一个豪华、典雅的商业空间，让每一个客户体会到尊贵、舒适的感觉。

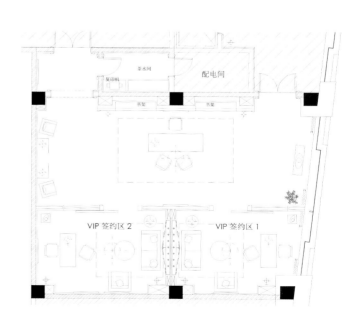

Shenyang Baoli Xinyu Garden Sales Office

沈阳保利心语花园售楼处

This case is up and down layers' monomer building. The first floor is for lobby, family exhibition area, negotiate area and bar. The second floor is for project cultural exhibition area, video projection room, signing room, VIP negotiation rooms and children activity area. The design's aim is to show noble elegant and make it become a modern design's model.

Among them, walk into the lobby, you can see an area of models hidden in the toughened glass ground. Look up, you can see it is surrounded by the golden corridors of the second floor. The lobby's design shows great temperament, luxurious and avant-garde, clearly showing building's design concept style.

设 计 师：王赟、王小峰
设计公司：广州尚逸装饰设计有限公司
建筑面积：1500平方米

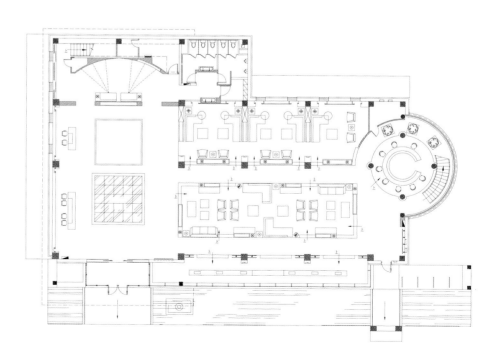

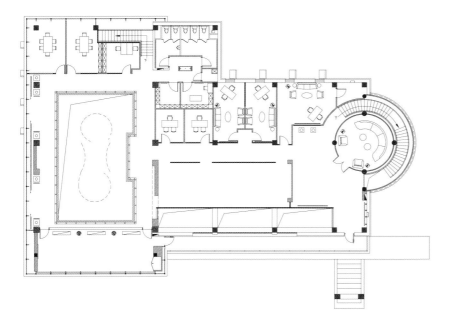

Stylist uses different design style in different areas, it can highlight the function of each area. The signing room is just one example. Signing room doesn't like the lobby's luxury feeling, it has more elegant comfortable atmosphere instead. Comfortable sofa, downy lamplight and pink walls create buyers a quiet comfortable environment to complete signing.

And sales department also designs a bar which makes events, parties' holding convenient. Light blue matches with white, which is quietly elegant and doesn't lose fashionable feeling. The design of beige leaves' droplights on the top surface add the space's gradation and details' adornment effect.

Estate sales center is not only for its commercial purpose to exist, but also becomes a high quality estate's image speak.

　　本案为上下两层单体建筑，首层为大堂、户型展示区、洽谈区、酒吧台，二层为项目文化展示区、视频放映室、签约室、VIP洽谈室、儿童活动区。设计旨在表现高贵优雅，使之成为现代设计的典范。

　　踏入大堂，可以看到一片区域的模型藏在钢化玻璃的地面，抬头就可以看到被二楼的金色回廊所包围。大堂设计表现得大气、豪华且前卫，明确地表现出楼盘的设计概念风格。

　　设计师在不同的区域运用不同的设计风格，更能突出每个区域的功能，而签约室就是其中的一个实例。签约室没有像大堂的豪华感觉，相反多了份优雅舒适的氛围。舒适的沙发、柔和的灯光，配合粉红色的墙身，为买家营造出一个安静、舒服的环境来完成签约。

　　售楼部还设计了一个酒吧台，方便举办活动、派对。浅蓝色与白色的配合，淡雅中不失时尚感。顶面米黄色叶子吊灯的设计增添了空间层次及细节的装饰效果。

　　楼盘销售中心不仅仅是为其商业目的而存在，同时也成为其高品质楼盘的形象代言。

Fuxing International Sales Offices in Wuhan City
武汉福星国际城售楼处

In this space, all the geometrical elements are perfectly fusing into each other but also stay their own independence. The strong comparison of three colors: black, white, red, derivates an unique space sensation. The over large shining ceiling strengthens the transportation space with a relatively lower height, and divides out different region functions. Meanwhile, the ceiling also provides for the whole sale building lighting resource, which models the modern gentle space theme. The conference room is installed with white titling illuminating pole that enlightens the black pole in the original architecture, breaking down the rigid space atmosphere.

设 计 师：李益中、周伟栋
设计公司：深圳市派尚环境艺术设计有限公司
主要建材：阻燃金属砂、深灰色氟碳漆、暗红色氟碳漆、玻璃丝印图案、白色哑光砖、深灰色哑光砖、黑色高亮砖

在这个空间里,各种几何元素既完美融合又各自独立,黑白红黄强烈的色彩对比碰撞出独特的空间感受。超大尺度的发光天花造型,以相对较低的标高,强化了交通空间,并分隔出不同的功能区域,同时也是整个销售中心重要的泛光源,塑造了柔和、现代的空间基调。洽谈区白色发光斜立柱的设置,与原建筑黑色直立柱交相辉映,打破了空间沉闷的气氛。

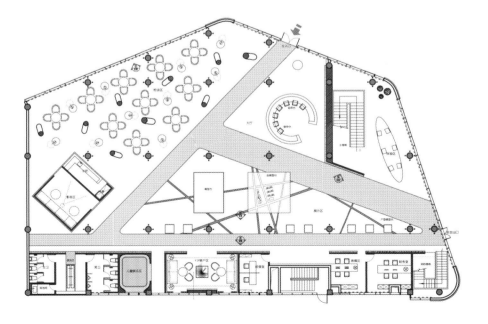

Zhonghai Jinsha Bay Sales Center

中海金沙湾东区售楼中心

The main part of reception desk in sale department adopts large black leather embedded with fake Swaroviski crystal grains, which forms the simple European stripes, then going with European stylish pole head and the pole body made of marble with black sand-blasting steel. Besides, we add lighting effect inside the marble that is under the reception desk, a modern approach demonstrating European style that more suits the taste of people nowadays.

The whole space is clearly directed by diamond shape signals; the large French widow in conference room fully explains the principle of "Customers come first"; the wine cabinet is sprayed with black piano varnish to have a light-penetrating disposal, which is echoing to the ceiling in the bar area and also mutual reflected by the light effect in the conference room. The whole prospect freely have one style of its own, highlighting the modern and luxurious tone.

设 计 师：周文胜、陆闻
设计公司：品格设计有限公司
项目地点：中国广东
建筑面积：650平方米
主要建材：沙漠风大理石、黑色镜钢、黑色扣皮、桔子玉大理石、透光片、黑白根木饰面

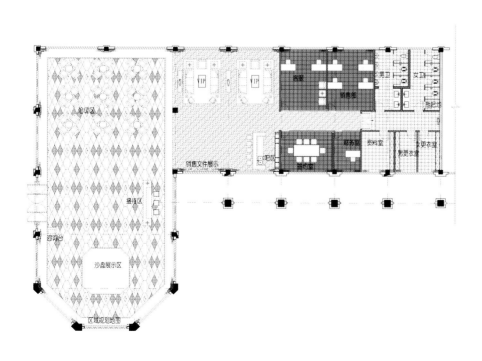

销售大厅接待台采用大面积的黑色皮镶嵌仿施华洛世奇水晶粒拼成的简欧风格花纹，配以简欧造型的柱头及由大理石和黑色镜钢喷砂造型所做的柱身，并在接待台下方大理石内部做打灯效果处理，利用现代的手法演绎出更符合现代人心理的欧式风情。

整个空间由菱形符号指引给人以明确导向性；洽谈区大面积的落地窗，充分注重"以客户为中心"的原则；酒柜则利用黑色钢琴漆边框做透光处理，与酒吧区天花及VIP洽谈区的透光效果相互辉映，自成一体，凸显其现代风尚与豪华格调。

Hangzhou Jindu Golf Sales Center

杭州金都高尔夫销售处

The Alps has amazingly wonderful beauty even in the different four seasons, thus the variety and color in the nival climate certainly is soul of the Alps. The design idea derives from these nature wanders of different mountain chains.

In the natural world, sunshine, sharp ice mountain, snow and moraine all have their own ecological traits even with more creativity and boundless vitality. Although they all grow into nature scene under the same circumstance, they have never appeared identical shape and figure.

The fusion of nature wanders into modern design, with the all sorts of flower decorations and rich architecture material, plus region division in levels, all of which create a natural marvelous sale center.

设 计 师：刘伟婷
设计公司：刘伟婷设计师有限公司
建筑面积：2800 平方米

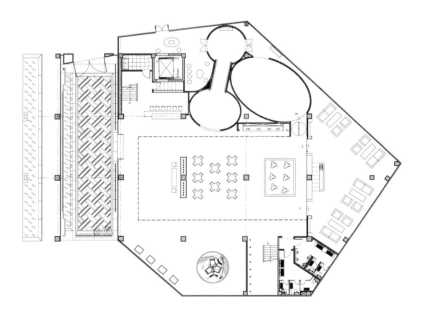

在阿尔卑斯山不同的季节都有美妙的丰富景色，而雪带气候变化和色彩就是阿尔卑斯山的灵魂。本案的设计意念就来自这些自然景观中不同山脉层次的奇观。

在自然界中，太阳的光线、冰山峰、冰雪、冰碛，这些山谷的自然美景，也有他们一套生态规律和性质，更是充满创造力和无限的生命力，虽然在同等条件下形成的自然景观，但从来没有出现相同的形式和线条。

把这些大自然的奇观奥妙溶入设计中，以不同形态的花饰、丰富的建筑物料和富有层次的区域划分，造就出一个带自然美感的销售空间。

Zhonghai Jincheng Club

中海锦城会所

The inspiration of the design in this case comes from the local history. This project is located in the first posthouse of the Southern Silk Road - Cuqiao, so the designer starts with Cuqiao's history. This place abounds with silk in the history and has a prosperous silk trading, therefore it is once known as the "Cocoon Bridge", and its glorious history is also derived from oval balls stacked by thousands upon thousands of silk - cocoon. Therefore, through the analysis and extraction of the cocoons' shape and characteristics, the designer gets the idea of unique theme patterns and use them in the interior design of the project, with the hope of bringing a beautiful prospect of pupating into a butterfly and beginning a prelude to the contemporary modern urban life here.

From the analysis of the composition in this case, the theme patterns are integrated into the new Art Deco design style, which make practical functions and aesthetics in one combined in this space. The stone materials, floor tiles, wallpaper, metallic paint, etc., which are with the visual elements full of modern sense and high quality, play the role of storytellers in the club - telling the stories of the local development history and the evolution of the city. Through the theme of the shadow of Cocoon, the design of this club tells you: "Who is the club? Where does the club come from? And where will the club go?"

设 计 师：方峻
设计公司：香港方黄建筑师事务所
项目地点：中国四川

由于该项目是在南方丝绸之路的第一个驿站——簇桥，于是，设计师追溯簇桥的历史渊源，该地在历史上盛产蚕丝，且蚕丝交易兴旺，曾一度被称为"茧桥"，其辉煌历史就是源自于千丝万缕叠积而成的椭圆形球体——"茧"。因此，设计师通过对"蚕茧"形态与特征的分析与提炼，形成独特的主题纹样，运用于该项目的室内设计之中，希望带给这里化蛹成蝶的美丽意境，在这里唱响当代现代化都市生活的序曲。

从设计构成分析看，把主题纹样图案融入新艺术装饰主义风格的空间之中，让空间集功能与审美于一体，用现代视觉元素和富有品质感的石材、地砖、墙纸、金属漆等材料在会所空间中扮演讲故事的人之角色，讲述当地历史发展与城市演变的故事。通过《"茧"影》这一主题，轻盈优雅地道出："会所是谁，会所从哪里来，会所要到哪里去！"

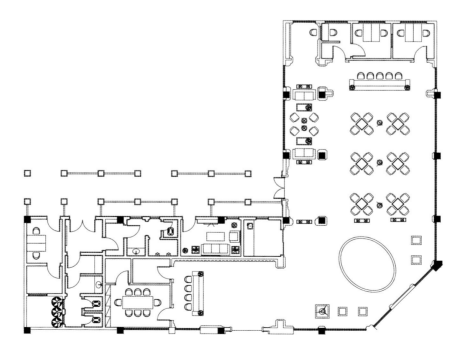

Huitong International Apartment Sales Department

汇通国际公寓售楼部

Since this case has limited area but enough storey height, we define the whole building with two floors. The first floor is designed to be functional service counter, sand box area and conference room. The second floor is divided into business conference room and office room. As to the choice of material, we adopt abundant wood, grey steel and stripe stone etc., so as to create a warm and shining atmosphere for the space. The stripe pattern on the theme wall becomes the theme of space, then we employ the mixing design of large amount of grey steel and stainless steel to make a decoration closet, which actually plays the role of screen, instead of being an simple ornament. The weaving grey steel on the side wall of conference room is echoing with the decoration closet and the screen on the second floor, which also turns out to be the main decoration in the conference room.

设计公司：J2/厚华顾问设计有限公司
项目地点：中国广东
建筑面积：200平方米
主要建材：泰柚、灰钢、不锈钢

　　本案由于平面面积小但是层高较高，所以设计时给空间设计定为两层。首层为功能服务台、沙盘区、洽谈区。二层为商业洽谈区和办公区。材料上采用了较多的木材和灰钢、木纹石等，主要打造出了空间的温馨感和闪亮感。主题墙木间条形成的图案成为了空间的主题。旁边采用大量灰钢和不锈钢交错设计的装饰柜，不但起到了装饰作用而且形成了屏风的功能。洽谈区侧墙体交错的灰色钢条呼应着装饰柜和二楼屏风，也成为洽谈区的主体装饰。

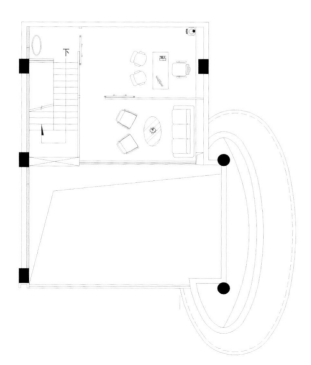
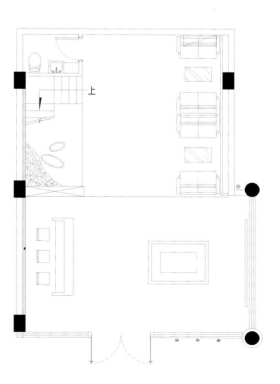

Pearl Park Sales Center
珠光新城销售中心

The plane plan for the whole sale department adopts unsymmetrical approach which breaks the rigid space relationship in many traditional sale centers. It creates an interestingly changeable space, with a perfect combination of wall straight line and beautiful figures, making the space more flexible and energetic. Naturally it will generate variable beauty of pattern and smooth gentle feeling in sensory organ.

When the client walks into the space that is pervaded with strong exotic gracefulness, he can enjoy a honorable guest treatment with every thoughtful detail. Wandering around the room, he will be easily grabbed by the structure of suspended ceiling that formed by six Roman poles. The special geometrical structure treats visitors with a grand visual banquet, which, it is said, is the stroke of a genius. The six poles already existed in the original space, but given that the highness of space is not enough, we make a structure like this without wasting. Walking around inside the house, people will be surprised everywhere. The related door arc, the Spanish architecture pole and crystal drop lamps, the charming county vision aroused by music, all of the scenery will be presented in front of clients , providing for them fully enjoyment.

After finishing the first floor area, the customer will come to a spot where is like " Jocob's ladder" in the Bible.

设 计 师：霍承显
设计公司：空间印象设计公司
项目地点：中国广东
建筑面积：2000平方米
主要建材：人造大理石、木、马赛克、窗帘布艺

People can rise themselves as they climb the stairs, and bright daylights sneak into the house through arc glass by the stair side, which enables people to enjoy the beautiful scenery. Going up further, visitors will enter the cozy resting room while talking to sale representatives. Several VIP rooms are built like wing-room also in Spanish style. The accessible material texture experience can offer sufficient texture touch, which 100 percent displays the honorable temperamental of the space. Beyond the reachable space, such the ceiling , uniquely made into a net structure of European wood inn, which cut down the cost but uncommon brand personality is still perfectly expressed...

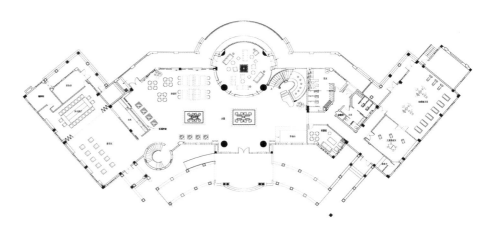

　　整个售楼中心的平面规划以非对称的手法，打破了多数售楼中心中规中矩的空间关系，营造了具有趣味性的空间变化，利用墙面直线与优美的曲线相结合，使空间更富有转折和张力，产生视觉上不断变化的形式美和感官上的流畅柔和。

　　当客户走进这个弥漫着浓浓的异国风情的空间中，每个贴心安排的细节，让他们感受着贵宾级的厚待。从容地漫步其中，视线被一个由六根罗马柱支撑起的米字形吊顶所吸引，特殊的几何造型让造访者霎那间感受到了气势不凡的视觉盛宴，据说这是出于设计师的神来之笔：六根柱子原是空间已有的架构，为了不造成新的浪费，考虑到一楼的空间层高偏低，故设计师做了这么一个造型。徜徉在室内，所经之处，高潮迭起：渐次连接的门拱，西班牙的建筑柱廊及水晶吊灯，音乐中透露的小镇风情——展现在眼前，让客户尽情地细意欣赏。

　　观摩过一楼的沙盘区后，客户经过一座宛如圣经故事上的"雅各布的天梯"，拾级而上，清透明亮的天光透过楼梯侧的拱门玻璃窗洒落，仿佛置身幸福国度。再往上，便可进入舒适的休憩区，并可随意与销售代表详谈。几间犹如厢房的贵宾室，同样是一番西班牙风情，随手可及的材质体验，质感十足，完美展现了空间的尊贵气派；而手能触及的范围之外，比如天花则被别有用心地做成了欧洲木屋的网架结构，成本降低了，非凡的品牌气质依然恰到好处……

Tianyi Bay Sales Office of Futong Real Estate
富通地产天邑湾售楼处

This case is located in Dongguan. Because the architecture always make most use of the circumstance and emphasize the nature growth structure, we agree to take good advantage of existed space (the upper space framework). As to the whole building style, we are in pursuit of the combination by tradition, nature and modern beauty appreciation, which will encourages the architecture to be fully presented. Therefore, on the choice of style, material, furniture, decoration and fabric, we all strictly follow the rule. The permeability of the space and the stretch to exterior scenery, thus, strengthen the privilege in visual field and improve the space psychological value.

设 计 师：韩松
设计公司：深圳市昊泽空间设计有限公司
项目地点：中国广东
建筑面积：620平方米

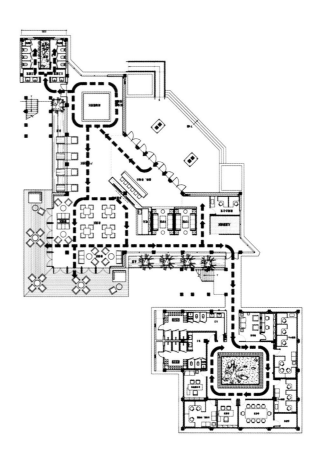

　　本项目位于东莞，由于建筑强势因势利导，强调自然生长结构，充分利用现有建筑空间的优势。在整体风格上，追求传统、自然和现代审美的融合，以使建筑之美完美地展现出来。因此在造型、材质、家具、饰品、面料的选择上都很好地遵循了这个原则。本案空间的通透性和对外部自然景观的引伸，则增加了空间视觉优势和空间心理价值。

Zhongxin Future City Sales Center

成都中信未来城售楼中心

The case is comprised by two parts - the left one and the right one, cut from the middle. They respectively provide with their own functions. In the first lobby, the logo " CITIC, THE FUTURE TOWN" immediately introduce people to a future space, and the blue curly lines design on the ceiling undoubtedly enhance the futurism for this space. The lighting decoration hanging on the ceiling form a real breathtaking prospect under the match of LED light effect of surrounding, which also implies the space theme. The other lobby provide a nice communicating spot for people who can sense elegance everywhere. Chinese passion and modern feeling are weaving around here.

设 计 师：邱春瑞、李赢
设计公司：台湾大易国际邱春瑞设计师事务所
项目地点：中国四川
建筑面积：950平方米
主要建材：爵士白、饰面板、山西黑、透光拉膜

本案整体上由两个大的部分组成，从中间分开，左右各为一处大的空间，这两处空间也应服务的需求提供不同的功能。一侧大厅中，"中信未来城"的标志将人们引入到一个未来的空间之中，天花上的蓝色曲线设计，无疑增强了这种空间的未来主义的特质。从天花上垂挂下来的灯饰，在周围LED灯光的配合下，美轮美奂，这也暗合了空间的主题意境。另一侧则为人们提供了一个很好的交流区。安静优雅的气息随处可见。中式的情怀与现代的感受在这里交织。

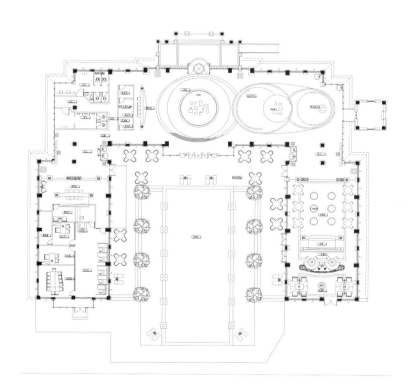

Zhongyin Uptown International Sales Center

中茵上城国际售楼部

Zhongyin Uptown International is a typical case of new classical style, located on the north side of Fuxia Road, around which there are convenient transportation and complete surrounding facilities. At the beginning of the design, the designer has done a research about the environment of the project. After a skillful combination of classical elements and casual but elegant modern techniques, the sales center reflects a distinct modern atmosphere and revives the essence of classicism.

设 计 师：周华美
设计公司：品川设计顾问有限公司
项目地点：中国福建
建筑面积：500平方米
主要建材：大理石、透光板、镜面、紫檀木、皮革、金钢板、银泊、不锈钢

Step into the 500-square-meter space, you will be greeted by exquisite furniture and comfortable atmosphere. Clever use of marble, leather, glass, panels, silver foil, stainless steel and mosaic fully displays the exquisiteness and nobleness of the space. As to the layout, the designer approached the ceiling with a simple and elegant style and placed a round suspended ceiling at the top of the model exhibition area. The huge dazzling round crystal chandelier is at the center of the space and becomes the visual focus. The two original square columns are changed into round columns and are embedded with carved patterns, which perfectly blend with the model exhibition area and the overall space, adding a splendid atmosphere into the space.

中茵上城国际是一个典型的新古典主义风格的空间，位于福州市福峡路北侧，交通便利，周边设施配套齐全。设计师在设计之初，就对环境进行考察，将古典元素和潇洒又精巧的现代手法融会贯通之后，为售楼部展现出鲜明的时代气息，同时复活了古典主义的精髓。

走进这个500平方米的空间，入眼的是精致的家具、舒适的环境。大理石、皮革、玻璃、面板、银箔、不锈钢、马赛克的巧妙运用，将精致且高贵的空间一一展现。在空间划分上，设计师将天花处理成简洁、高雅的格调，并在沙盘的顶上安置了一个圆形吊顶，璀璨的巨型圆形水晶灯在空间的中间，成为空间的视觉焦点。将原本的两根方柱改为圆柱，再嵌上雕花，与沙盘及整体空间完美结合，也为空间增添了华丽的气质。

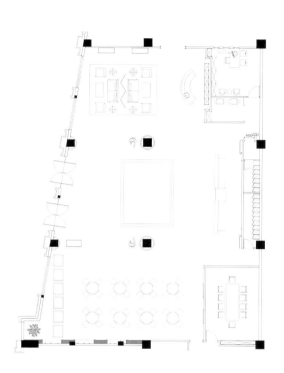

Shenyang Yameilijia Sales Center

沈阳亚美利加售楼中心

The overall background of the design in the case adopts the dark colour, using black and white to express the noble, restrained and fashionable style of the space. The interlacement of black and white brings with a powerful tension in a clear way and a coexistence of harmony and stableness. The design of the strip on the top extends and stretches the visual, increasing the sense of space. And flexible decoration gives a amiable and fresh feeling. The tubes hanging around the column distribute the soft lighting, adding a sense of fantasy and elegance to the interior, which is a major highlight of the room.

设 计 师：孙大为
设计公司：沈阳点石（国际）室内设计顾问有限公司
项目地点：中国辽宁
建筑面积：800平方米
主要建材：黑镜、黑钛钢、凯撒灰石材、发光软膜

本案设计整体背景采用暗色处理，利用黑白两色来表达空间的高贵、矜持与时尚格调，两种颜色相互交错，以一种清晰的方式给人们带来强大的张力，和谐与稳重并存。顶部的长条形设计使视觉得以延伸和舒展，增大了空间感。软装给人亲切清晰的感觉，围绕圆柱垂吊下的灯管散发着柔和的灯光，为室内增添了虚幻与高贵感，成为室内的一大亮点。

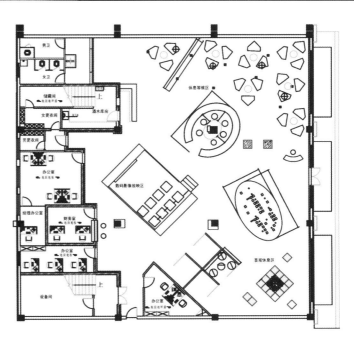

图书在版编目（CIP）数据

群菁汇·高端售楼会所大赏 / 徐宾宾 编. —南京：江苏人民出版社，2011.8
ISBN 978-7-214-05838-6
Ⅰ. ①群… Ⅱ. ①徐… Ⅲ. ①商业建筑－建筑设计
－作品集－世界②休闲娱乐－服务建筑－建筑设计－作品集－世界 Ⅳ. ①TU247

中国版本图书馆CIP数据核字(2011)第131158号

群菁汇·高端售楼会所大赏

徐宾宾 编

责任编辑	刘焱　张蕊
责任监印	马琳
出　　版	江苏人民出版社（南京湖南路1号A楼）
发　　行	天津凤凰空间文化传媒有限公司
销售电话	022-87893668
网　　址	http://www.ifengspace.cn
集团地址	凤凰出版传媒集团（南京湖南路1号A楼　邮编：210009）
经　　销	全国新华书店
印　　刷	当纳利（深圳）印刷有限公司
开　　本	965毫米×1270毫米　1/16
印　　张	23
字　　数	184千字
版　　次	2011年8月第1版
印　　次	2011年8月第1次印刷
书　　号	978-7-214-05838-6
定　　价	328.00(USD 58.00)

（本书若有印装质量问题，请向发行公司调换）